STRUCTURAL ANALYSIS SYSTEMS

Software — Hardware
Capability — Compatibility — Applications

Volume 2

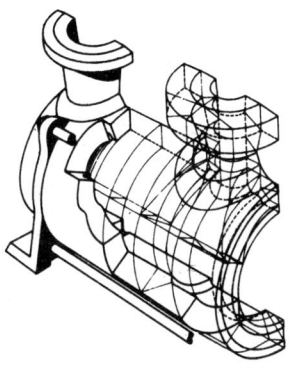

An international series of practical guidebooks
on structural analysis systems and their applications

Other Pergamon Titles of Interest

BATHE	Nonlinear Finite Element Analysis and ADINA 1983
COHN & MAIER	Engineering Plasticity by Mathematical Programming
COWAN	Predictive Methods for the Energy Conserving Design of Buildings
CROUCH	Matrix Methods Applied to Engineering Rigid Body Mechanics
GIBSON	Thin Shells
HARRISON	Structural Analysis and Design
HEARN	Mechanics of Materials, 2nd Edition
HOLLAND	Microcomputers and Their Interfacing
HORNE	Plastic Theory of Structures
JAMSHIDI & MALEK-ZAVAREI	Linear Control Systems
LEININGER	Computer Aided Design of Multivariable Technological Systems
LIVESLEY	Matrix Methods of Structural Analysis, 2nd Edition
NOOR & HOUSNER	Advances and Trends in Structural and Solid Mechanics
NOOR & McCOMB	Computational Methods in Nonlinear Structural and Solid Mechanics
PARKES	Braced Frameworks, 2nd Edition
ROZVANY	Optimal Design of Flexural Systems
SPILLERS	Automated Structural Analysis
WARBURTON	Dynamical Behaviour of Structures, 2nd Edition

Pergamon Related Journals *(Free Sample Copy Gladly Sent on Request)*

BUILDING AND ENVIRONMENT

CEMENT AND CONTRETE RESEARCH

CIVIL ENGINEERING FOR PRACTICING AND DESIGN ENGINEERS

COMPUTERS AND GRAPHICS

COMPUTERS AND INDUSTRIAL ENGINEERING

COMPUTERS AND STRUCTURES

FATIGUE AND FRACTURE OF ENGINEERING MATERIALS AND STRUCTURES

INTERNATIONAL JOURNAL OF APPLIED ENGINEERING EDUCATION

INTERNATIONAL JOURNAL OF SOLIDS AND STRUCTURES

JOURNAL OF ENGINEERING AND APPLIED SCIENCES

MATHEMATICAL MODELLING

STRUCTURAL ANALYSIS SYSTEMS

Software — Hardware
Capability — Compatibility — Applications

A. NIKU-LARI
Director, Institute for Industrial Technology Transfer
24 Rue des Mimosas, Gournay s/Marne
F93460 France

Volume 2

PERGAMON PRESS
OXFORD · NEW YORK · TORONTO · SYDNEY · FRANKFURT

U.K.	Pergamon Press Ltd., Headington Hill Hall, Oxford OX3 0BW, England
U.S.A.	Pergamon Press Inc., Maxwell House, Fairview Park, Elmsford, New York 10523, U.S.A.
CANADA	Pergamon Press Canada Ltd., Suite 104, 150 Consumers Road, Willowdale, Ontario M2J 1P9, Canada
AUSTRALIA	Pergamon Press (Aust.) Pty. Ltd., P.O. Box 544, Potts Point, N.S.W. 2011, Australia
FEDERAL REPUBLIC OF GERMANY	Pergamon Press GmbH, Hammerweg 6, D-6242 Kronberg, Federal Republic of Germany
JAPAN	Pergamon Press Ltd., 8th Floor, Matsuoka Central Building, 1-7-1 Nishishinjuku, Shinjuku-ku, Tokyo 160, Japan
BRAZIL	Pergamon Editora Ltda., Rua Eça de Queiros, 346, CEP 04011, São Paulo, Brazil
PEOPLE'S REPUBLIC OF CHINA	Pergamon Press, Qianmen Hotel, Beijing, People's Republic of China

Copyright © 1986 Pergamon Press Ltd.

All Rights Reserved. No part of this publication may be reproduced, stored in a retrieval system or transmitted in any form or by any means: electronic, electrostatic, magnetic tape, mechanical, photocopying, recording or otherwise, without permission in writing from the publishers.

First edition 1986

Library of Congress Cataloging in Publication Data
Structural analysis systems.
Includes indexes.
1. Structures, Theory of—Data processing—
Addresses, essays, lectures. I. Niku-Lari, A.
TA647.S77   1985   624 1'7'0285   85-9419

British Library Cataloguing in Publication Data
Niku-Lari, A.
Structural analysis systems: software, hardware, capability, compatibility, applications.
Vol. 2.
1. Structures, Theory of—Data processing
I. Title
624.1'71'02854   TA647
ISBN 0-08-032578-5

Cover drawing: Centrifugal pump casing.
Manufacturer: C.C.M.-Sulzer, France.
Software used: CA.ST.OR

Printed in Great Britain by A. Wheaton & Co. Ltd., Exeter

INTERNATIONAL EDITORIAL ADVISORY COMMITTEE

Dr. T. ANDERSSON, *Sweden*
Prof. J. H. ARGYRIS, *Federal Republic of Germany*
Prof. K. J. BATHE, *USA*
Prof. T. BELYTSCHKO, *USA*
Dr. M. BERNADOU, *France*
Prof. B. A. BILBY, *UK*
Dr. A. CHAUDOUET, *France*
Prof. R. D. COOK, *USA*
Dr. T. FUTAGAMI, *Japan*
Prof. GUO YOUZHONG, *People's Republic of China*
Dr. L. IMRE, *Hungary*
Prof. H. LIEBOWITZ, *USA*
Mr. J. MACKERLE, *Sweden*
Mr. W. M. MAIR, *UK*
Dr. G. A. MILIAN, *Mexico*
Dr. D. NARDINI, *Yugoslavia*
Dr. I. PACZELT, *Hungary*
Prof. G. SANDER, *Belgium*
Prof. R. P. SHAW, *USA*
Prof. M. TANAKA, *Japan*
Prof. W. N. WENDLAND, *Federal Republic of Germany*

PREFACE

Recent years have seen a rapid increase in the number of structural analysis software existing in the world market.

Most of the current software journals are based on theoretical background. They provide academics and software developers with a very useful tool. People from industry however, who are non-specialists in finite or boundary element methods, have often difficulty in finding industry-oriented documents to help them select the software and hardware most suited to their needs.

This guidebook series aims to provide the engineer with up-to-date information about structural analysis systems existing around the world.

Each paper gives detailed information about a specific software, its capability, its limitations and several practical examples from industry with computer and user cost. It also gives to the user the necessary information about postprocessor capabilities, computer-aided design connection and software compatibility with the most common computers.

Most papers published in this volume follow the same logical structure to allow interactive comparison.

Our concern is to promote international co-operation on this important subject and to contribute to a better understanding between research and industry.

I would like to thank the distinguished members of the editorial committee for their scientific and technical help which made the publication of the present volume possible.

A Niku-Lari
Editor

EDITORIAL

Structural analysis aims to construct numerical models which represent the best behaviour of the actual engineering material and component. These models are used in research for better understanding of experimental results. In industry the structural analysis models allow both the optimization of design and the prediction of failure.

The structural analysis (SA) is therefore a multidisciplinary problem which demands knowledge of several scientific and industrial disciplines such as, engineering sciences, mechanical or civil engineering, informatics, applied mathematics, computer sciences, etc.

International competition gives to industry the necessary impulse to optimise the design of parts and structures.

The engineer should save material and energy and use new and lighter materials such as composites. No longer is one allowed to over-design parts for "security reasons", and new international criteria have to be considered.

Industry needs to design sophisticated parts working in very special environments, in space, in the human body, in the sea, etc.

The compatibility of the structural analysis systems with modern micro-computers allow small and medium size companies to make use of these new technologies. New super computers help to find rapid solutions to complex industrial design problems.

The evolution of interactive graphics allow the full integration of structural analysis programs in a computer aided design and manufacturing environment. Expert systems, application of artificial intelligence and computer-aided decision making bring new developments in this field.

Structural analysis systems existing in the world market are powerful bridges between research and industry. They bring theory in direct physical contact with the industrial application.

The SA technology is in a rapid evolution. More and more new computers and powerful software appear in the market and the industry faces a new problem – that of selecting the optimum structural analysis system.

The choice of a structural analysis software is an important decision which can often exercise a significant influence over the successful development of a research, manufacture, or design project. Depending on the engineering problem and hardware available, a good choice of the computer program can be both cost and time effective. This new international guidebooks series aims to provide the engineer with the most up-to-date information about structural analysis systems currently available in the world market, and their capabilities.

Editorial

Published under the guidance of a distinguished scientific committee whose members are internationally recognised specialists of finite or boundary element methods, the series should be considered as an essential practical reference tool for the modern engineer involved in such areas as structural, mechanical, civil, nuclear, aeronautical and design engineering, computer science and software development.

Each volume gives detailed information about a wide range of selected software packages describing their purpose, capabilities and limitations and provides several practical examples of industrial applications, often supported by case studies. It also gives to the user the necessary information about postprocessor capabilities, computer-aided design integration and software compatibility with the most commonly used computers.

The guidebooks are industry-oriented and should prove indispensible in helping potential users to select the soft and hardware most suited to their needs. Each volume commences with a program description in tabular form, rapidly directing readers to the program most likely to solve their industrial problems, and concludes with a case-study index.

Main areas covered in the series:

- Finite and boundary element programs
- Finite difference and other methods
- Computer graphics
- Artificial intelligence and expert systems
- Computer-aided decision making in engineering
- Computer-aided design and manufacturing (CAD/CAM)
- Integration of structural analysis and expert systems in engineering CAD/CAM environment
- Hard and software selection
- Micro-computer applications in engineering
- New development in structural analysis software and interactive graphics
- Industrial case study
  . Mechanical engineering
  . Aeronautics and nuclear
  . Biomechanics
  . New materials (composites, plastics, etc)
  . Civil engineering (offshore, seismic, earthquake, etc).

Authors wishing to submit a paper under one of the above headings for possible publication in future volumes are invited to submit their manuscript for editorial consideration of the international scientific committee to the address below.

Dr A Niku-Lari, Director
Institute for Industrial Technology Transfer
I.I.T.T.-international
24 Rue des Mimosas
93460 Gournay-sur-Marne
FRANCE
TEL: (1) 43.05.17.19

CONTENTS

Program Description Tables	xiii
AFAG: A Precise Linear and Nonlinear Analysis of Arches and Similar Structures by Means of the Flexibility Method	1
L. Cappellari and R. Gori	
AQUADYN: A Linear Modelling Tool in Hydrodynamics	9
J. P. Le Fèvre	
ASE: Analysis of Space Structures	15
C. Katz-Fink Stieda	
BEMFFT: A Program for the Numerical Solution of One-dimensional Boundary Integral Equations	19
U. Lamp	
BOSOR4: Program for Stress, Stability and Vibration of Complex Branched Shells of Revolution	25
D. Bushnell	
BOSOR5: Program for Buckling of Complex, Branched Shells of Revolution including Large Deflections, Plasticity and Creep	55
D. Bushnell	
ESA: Engineering Structural Analysis on Personal Computers	69
J. P. Rammant	
HYBRID: Determination of Notch and Crack Boundary Stresses using Hybrid Elements with the Program-System HYBRID	79
Drumini, Rudi and Carmine	
INFESA: Interactive Finite Element Stress Analysis	87
A. O. Moscardini and B. A. Lewis	

MEF/MOSAIC: Linear, Nonlinear Finite Element Code with Interactive
Graphic Display 95
  J. F. Cochet

RCAFAG: On the Stability Analysis of Reinforced Concrete Arches
and Frames in the Geometrical and Nonlinearities Field 113
  L. Cappellari and R. Gori

REST: A Computer Program for Response Statistics of Discretized
Structures to Random Excitations 119
  C. W. S. To

SIMP: Design Sensitivity, Advanced CAE and CAT 129
  K. Zimmermann

STAR2: Static Analysis of Spatial Frame Structures 145
  C. Katz-Fink-Stieda

STRUGEN: Pre- and Postprocessing with STRUGEN, Coloured Interpretation
of Results of Stress Output 147
  J. Simon-Weidner

Y12M: Solving Large Systems of Linear Algebraic Equations by the Use
of Package Y12M 151
  Z. Zlatev

Finite Element Software for Stress Analysis of Laminated Composites 157
  J. Mackerle

Case Study Index 221

PROGRAM
DESCRIPTION
TABLES

Program Description Tables

	METHOD				ELEMENT LIBRARY							GEOMETRY			SEE VOL
	Finite element	Boundary element	Finite difference	Other	Truss/beams	2D membranes	Plates	Shells	Axisymmetric	3D solids	Boundary elements	Special elements	2D analysis	3D analysis	Axisymmetric	
ADINA	x				x	x	x	x	x	x			x	x	x	1
AFAG	x			x	x									x		2
AIT	x				x	x	x	x	x				x	x	x	1
ALSA	x				x	x	x	x					x	x		3
ANSYS	x				x	x	x	x	x	x	x	x	x	x	x	1
AQUADYN		x		x						x				x		2
ASE	x				x	x	x	x	x	x	x	x	x	x	x	2
AXISYMMETRIC	x							x	x		x				x	1
BEASY		x									x		x	x	x	1
BEFE	x	x			x	x	x	x		x	x	x	x	x		3
BEMFFT		x		x							x		x			2
BEWAVE		x									x	x	x	x		3
BOSOR4	x		x			x	x						x		x	2
BOSOR5	x		x			x	x						x		x	3
CASTEM	x				x	x	x	x	x	x		x	x	x	x	3
CASTOR	x	x			x	x	x	x	x	x	x	x	x	x	x	1
DAPST	x				x	x	x	x	x				x		x	1
DEFOR	x				x									x		1
ELASTODYNAMICS (2D)		x									x	x	x			2
ESA	x				x	x	x	x	x			x	x	x	x	2
FEMFAM	x				x	x	x	x	x	x		x	x	x	x	3
FEMPAC	x				x	x	x	x	x	x			x	x	x	3
FENRIS	x				x	x	x	x	x	x		x	x	x	x	3
FIESTA	x									x				x		3
FLASH	x				x		x	x	x			x		x	x	1
FLEXAN	x										x			x		3
HYBRID	x					x						x	x			2
IBA	x				x	x	x	x	x				x	x	x	1
INFESA	x					x							x			2

TABLE 1. Modelization and Type of Discretization.

Program Description Tables

	METHOD				ELEMENT LIBRARY								GEOMETRY			SEE VOL
	Finite element	Boundary element	Finite difference	Other	Truss/beams	2D membranes	Plates	Shells	Axisymmetric	3D solids	Boundary elements	Special elements	2D analysis	3D analysis	Axisymmetric	
KYOKAI	x	x				x			x	x	x	x	x	x	x	1
LASSAQ	x				x		x	x					x			3
MEF/MOSAIC	x				x	x	x	x	x	x		x	x	x	x	2
MICRO STRESS	x				x								x	x		1
MODULEF	x				x	x	x	x	x	x		x	x	x	x	3
MSRC-RB	x								x					x		3
NE XX	x				x					x				x		1
OSTIN		x								x			x		x	3
PAFEC	x	x			x	x	x	x	x	x	x	x	x	x	x	1
PAID	x				x							x		x		1
PANDA			x										x			1
PDA/PATRAN	x			x	x	x	x	x	x	x			x	x	x	2
RAPS	x				x	x	x	x	x	x		x	x	x	x	3
RCAFAG	x			x	x								x			2
REST	x				x	x	x	x	x	x			x	x	x	2
ROBOT	x					x	x	x					x		x	3
S AND CM	x				x	x		x				x	x	x		1
SAMKE	x				x	x	x	x	x	x		x	x	x	x	3
SESAM '80	x				x	x	x	x	x	x		x	x	x	x	1
SIMP	x		x	x	x	x	x	x					x	x		2
STAR 2	x				x					x	x			x		2
STDYNL	x	x			x	x	x	x		x			x			3
STRUGEN				x	x	x	x	x		x			x	x		2
SURFOPT	x					x		x					x		x	3
THERMAL	x				x	x							x	x		1
TITUS	x				x	x	x	x	x	x		x	x	x	x	1
UCIN GEAR	x				x								x			1
Y1 2M																2
ZERO-4	x					x		x	x	x	x	x	x	x	x	3

TABLE 1. Modelization and Type of Discretization (continued).

Program Description Tables xvii

	MATERIAL								CAPABILITIES							SEE VOL
	Linear elastic isotropic	Linear elastic anisotropic	Elasto-plastic	Nonlinear elastic	Viscoelastic/creep	Composites	Soil	Concrete	Static analysis	Dynamic analysis	Geometric nonlinear	Buckling/postbuckling	Heat transfer	Fracture mechanics	Fluid/structure inter	
ADINA	x	x	x	x	x	x	x	x	x	x	x	x	x	x	x	1
AFAG	x								x							2
AIT	x								x		x				x	1
ALSA	x	x							x	x						3
ANSYS	x	x	x	x	x	x		x	x	x	x	x	x	x	x	1
AQUADYN										x					x	2
ASE		x							x	x						2
AXISYMMETRIC	x								x	x	x	x	x		x	1
BEASY	x	x							x				x	x		1
BEFE	x	x				x			x							3
BEMFFT	x								x							2
BEWAVE					x					x				x		3
BOSOR4	x	x			x				x	x	x	x				2
BOSOR5	x	x	x		x	x			x		x	x				3
CASTEM	x	x	x	x	x	x	x	x	x	x	x	x	x	x	x	3
CASTOR	x	x			x				x	x		x	x	x	x	1
DAPST	x	x			x				x	x						1
DEFOR	x	x							x		x					1
ELASTODYNAMICS (2D)	x									x						2
ESA	x	x		x					x	x			x			2
FEMFAM	x	x							x	x	x	x	x		x	3
FEMPAC	x	x							x	x			x			3
FENRIS	x		x	x					x	x	x	x			x	3
FIESTA	x	x							x				x	x		3
FLASH	x	x							x			x				1
FLEXAN		x							x	x					x	3
HYBRID	x								x							2
IBA	x	x			x	x			x	x			x			1
INFESA	x				x				x							2

TABLE 2. Materials and Analysis Capabilities.

Program Description Tables

	MATERIAL							CAPABILITIES							SEE VOL
	Linear elastic isotropic	Linear elastic anisotropic	Elasto-plastic	Nonlinear elastic	Viscoelastic/creep	Composites	Soil	Concrete	Static analysis	Dynamic analysis	Geometric nonlinear	Buckling/postbuckling	Heat transfer	Fracture mechanics	Fluid/structure inter	
KYOKAI	x	x			x	x			x				x			1
LASSAQ		x				x			x							3
MEF/MOSAIC	x	x		x		x			x	x	x	x	x	x		2
MICRO STRESS	x								x							1
MODULEF	x	x	x	x		x			x	x	x		x			3
MSRC-RB		x	x						x							3
NE XX	x								x							1
OSTIN				x		x			x							3
PAFEC	x	x	x	x	x	x	x		x	x	x	x	x	x	x	1
PAID	x								x	x			x		x	1
PANDA	x	x	x			x			x			x				1
PDA/PATRAN	x	x							x							2
RAPS									x	x			x	x		3
RCAFAG	x			x	x	x			x							2
REST	x									x						2
ROBOT	x	x							x							3
S AND CM	x								x	x		x				1
SAMKE	x	x				x			x		x		x			3
SESAM '80	x	x	x				x		x	x				x	x	1
SIMP	x								x	x						2
STAR 2				x		x		x	x							2
STDYNL	x			x			x		x	x	x					3
STRUGEN																2
SURFOPT	x								x							3
THERMAL	x								x							1
TITUS	x	x	x	x	x		x		x	x	x	x	x	x	x	1
UCIN GEAR	x								x							1
Y12M																2
ZERO-4	x								x	x					x	3

TABLE 2. Materials and Analysis Capabilities (continued).

Program Description Tables

	LOADING								PRE/POSTPROCESSING							SEE VOL
	Nodal/line	Pressure	Selfweight	Centrifugal	Thermal	Heat flux	Prescribed displacement	Other	Free format input	Mesh generation	Plot routines	Automatic node number	Combinations of load cs	Interactive graph	CAD interfaces	
ADINA	x	x	x	x	x	x	x	x	x	x	x	x	x	x	x	1
AFAG	x	x	x		x		x	x	x	x	x	x	x	x		2
AIT	x	x	x		x		x		x							1
ALSA	x	x	x				x	x		x	x	x	x	x		3
ANSYS	x	x	x	x	x	x	x	x	x	x	x	x	x	x	x	1
AQUADYN								x	x	x	x	x		x		2
ASE	x	x	x		x		x		x	x	x	x	x			2
AXISYMMETRIC	x	x						x					x			1
BEASY	x	x	x	x	x	x	x	x	x	x	x	x	x	x	x	1
BEFE	x	x	x	x			x	x		x	x		x	x		3
BEMFFT							x				x					2
BEWAVE		x												x	x	3
BOSOR4	x	x		x	x		x	x	x	x	x	x				2
BOSOR5		x		x			x	x	x	x	x	x				3
CASTEM	x	x	x	x	x	x	x	x	x	x	x	x	x	x		3
CASTOR	x	x	x	x	x	x	x	x	x	x	x	x	x	x	x	1
DAPST	x	x	x				x		x		x	x				1
DEFOR	x		x		x		x	x	x		x		x	x		1
ELASTODYNAMICS (2D)	x						x	x			x					2
ESA	x	x	x	x	x	x	x	x	x	x	x	x	x	x	x	2
FEMFAM	x	x	x	x	x	x	x	x	x	x	x	x	x	x		3
FEMPAC	x	x	x	x	x	x	x	x	x	x	x	x	x	x	x	3
FENRIS	x	x	x		x		x	x	x	x	x	x	x	x		3
FIESTA	x	x	x	x	x	x	x	x	x	x	x	x	x			3
FLASH	x	x	x	x	x		x	x	x	x	x	x	x	x	x	1
FLEXAN	x		x		x		x	x	x	x	x			x		3
HYBRID	x						x		x	x	x	x				2
IBA	x	x	x		x		x	x	x	x	x	x	x	x		1
INFESA	x	x					x		x	x	x	x		x		2

TABLE 3. Loadings and User Comfort.

Program Description Tables

	LOADING							PRE/POSTPROCESSING							SEE VOL
	Nodal/line	Pressure	Selfweight	Centrifugal	Thermal	Heat flux	Prescribed displacement	Other	Free format input	Mesh generation	Plot routines	Automatic node number	Combinations of load cs	Interactive graph	CAD interfaces	
KYOKAI	x	x			x	x	x	x	x	x	x	x		x		1
LASSAQ	x															3
MEF/MOSAIC		x	x	x	x	x	x	x	x	x	x	x		x		2
MICRO STRESS	x	x	x		x		x		x		x					1
MODULEF	x	x	x	x	x	x	x	x	x	x	x	x		x		3
MSRC-RB	x	x	x				x	x	x				x			3
NE XX	x		x				x	x	x			x	x	x		1
OSTIN								x		x			x			3
PAFEC	x	x	x	x	x	x	x	x	x	x	x	x	x	x	x	1
PAID	x	x	x		x	x	x	x			x		x	x		1
PANDA							x		x			x				1
PDA/PATRAN	x	x	x		x		x	x	x	x	x	x	x	x	x	2
RAPS									x		x		x	x		3
RCAFAG	x	x	x		x		x	x	x	x	x	x		x		2
REST							x			x						2
ROBOT	x	x	x	x	x		x	x				x	x			3
S AND CM	x	x										x	x			1
SAMKE	x	x	x		x		x	x	x	x	x	x	x	x	x	3
SESAM '80	x	x	x	x	x		x	x	x	x	x	x	x	x	x	1
SIMP	x			x		x	x	x	x	x	x	x	x	x	x	2
STAR 2	x	x	x		x		x	x	x	x	x	x	x			2
STDYNL	x	x	x		x		x	x	x	x		x	x			3
STRUGEN									x	x	x	x		x		2
SURFOPT	x						x		x	x	x	x	x			3
THERMAL	x	x			x							x	x			1
TITUS	x	x	x	x	x	x	x	x	x	x	x	x	x	x	x	1
UCIN GEAR		x							x	x	x	x				1
Y1 2M																2
ZERO-4	x	x	x		x		x			x	x	x		x		3

TABLE 3. Loadings and User Comfort (continued).

Program Description Tables

	HARDWARE													SEE VOL
	CDC	IBM	Univac	Cray	Amdahl	Honeywell	Data General	Prime	VAX, DEC	HP	Apollo	Microcomputers	Other mainframes	Other minicomputers	
ADINA	x	x	x	x		x	x	x	x			x	x	x	1
AFAG										x					2
AIT	x	x								x					1
ALSA		x		x									x		3
ANSYS	x	x	x	x	x		x	x	x	x	x		x	x	1
AQUADYN				x					x						2
ASE	x	x								x					2
AXISYMMETRIC										x					1
BEASY	x	x	x	x				x		x			x	x	1
BEFE		x						x	x						3
BEMFFT								x							2
BEWAVE	x								x					x	3
BOSOR4	x	x	x						x						2
BOSOR5	x	x	x						x						3
CASTEM	x	x	x	x				x	x					x	3
CASTOR	x	x	x	x					x						1
DAPST	x									x					1
DEFOR		x							x	x				x	1
ELASTODYNAMICS (2D)		x											x		2
ESA												x		x	2
FEMFAM								x							3
FEMPAC	x	x	x		x	x	x	x	x	x	x	x	x	x	3
FENRIS		x		x					x		x			x	3
FIESTA	x								x						3
FLASH	x	x	x			x	x	x		x			x	x	1
FLEXAN	x	x		x					x						3
HYBRID												x			2
IBA		x							x	x				x	1
INFESA									x					x	2

TABLE 4. Hardware Compatibilities.

Program Description Tables

	HARDWARE														SEE VOL
	CDC	IBM	Univac	Cray	Amdahl	Honeywell	Data General	Prime	VAX, DEC	HP	Apollo	Microcomputers	Other mainframes	Other minicomputers	
KYOKAI									x	x					1
LASSAQ									x						3
MEF/MOSAIC	x	x	x			x	x	x	x						2
MICRO STRESS												x			1
MODULEF	x	x	x	x		x			x		x		x		3
MSRC-RB													x		3
NE XX											x				1
OSTIN									x	x					3
PAFEC	x	x	x	x	x	x	x	x	x	x	x		x	x	1
PAID	x						x								1
PANDA	x	x					x								1
PDA/PATRAN	x			x		x	x	x			x		x	x	2
RAPS		x							x	x	x				3
RCAFAG										x					2
REST	x														2
ROBOT	x												x		3
S AND CM												x			1
SAMKE	x	x	x				x								3
SESAM '80	x	x					x	x						x	1
SIMP	x		x	x			x								2
STAR 2	x	x									x				2
STDYNL	x	x											x		3
STRUGEN	x	x							x				x		2
SURFOPT			x												3
THERMAL												x			1
TITUS	x	x	x						x		x				1
UCIN GEAR		x		x											1
Y12M	x	x	x												2
ZERO-4		x												x	3

TABLE 4. Hardware Compatibilities (continued).

AFAG: A PRECISE LINEAR AND NONLINEAR ANALYSIS OF ARCHES AND SIMILAR STRUCTURES BY MEANS OF THE FLEXIBILITY METHOD

L. Cappellari and R. Gori

Istituto di Scienza delle costruzioni, Università di Padova, Italy

ABSTRACT

AFAG is a dual finite element programme for structural analysis based on the flexibility method.

It permits static, linear and nonlinear structural analysis of deformable elastic structures such as arches and similar structures (bridge arches, portal frames, tubes, dynamometric rings, etc.).

The input is extremely articulated and allows the implementation of diverse external loads and prestrains conditions as cause of stress (shrinkage, creep, thermal variations, inelastic constraint displacements).

A programme subroutine calculates the influence lines of the internal forces and of the components of displacement of the structural sections.

Prints and/or graphics of the internal forces of single loading, or of a combination of load conditions and of the relative deformed shapes and of the influence lines can be requested in output.

THEORETICAL BACKGROUND

The flexibility method has been used to give a drastic reduction in the number of unknowns, while the displacements are determined directly with the virtual force theorem. This has enabled the development of a programme for desktop or micro computer.

The flexibility coefficients of the elasticity equations are obtained, element by element, by integrating the products of linear or parabolic force functions along beam elements with linearly variable cross sections.

The knowns are obtained in the same way.

The same algorithms used for the determination of the unitary equilibrated states on the statically determined systems can be used for the evaluation of the displacements u, v and φ by means of the virtual force theorem.

The procedure of the second order analysis involves the automatic consideration, at each step, of the new structure coinciding with the deformed shape of the preceding step until convergence is obtained, according to an iterative process.

It should be stressed that the programme takes into account deformability due to bending moment, axial force and shear force.

FIELD OF APPLICATION

- Geometrical. The analysis concerns plane systems made up of variable sections, curved axis beams.

- Materials. Linear elastic isotropic material is considered in every analysis.

- Analysis capabilities. The programme executes first and second order static analysis.

- Loadings. Selfweight, external loads (point and line loads), residual stresses (thermal loadings, distortions, prestrains, creep, shrinkage).

PROGRAM DESCRIPTION

- Method: finite element flexibility method.

- Type of elements: beams.

- Programme structure:
    user comfort: . user facilities: data input from keyboard, or partial or total reuse of data from previous executions; interactive use; direct exploitation of results (forces and displacements) or storage in files for later use.

    . automatic mesh generation and node and element numbering.

    . post processor: drawing of the results, stress calculation, reaction calculation, combination of loading cases.

    R & D: . language of the program: Basic.

    . subprogrammes and their interrelations: see flow-chart (see Table I and Fig. 1)

HARDWARE COMPATIBILITIES

- Minimal configuration of materials required: 24 K RAM computer, printer and plotter.

- Type of computer and peripherals: Hewlett-Packard.

- Media: available in magnetic tape.

EXAMPLES OF APPLICATION

Steel Portal Frame for Industrial Construction. Second Order Analysis

The structure analyzed here is a steel portal frame fixed at the bases, designed for industrial buildings. The frame is made up of I PB 240 DIN 1025 St 37-2 (EURONORN 53-62 (HE-B)) irons. Given the dimensions of the beams and the forces

AFAG: A Precise Linear and Nonlinear Analysis

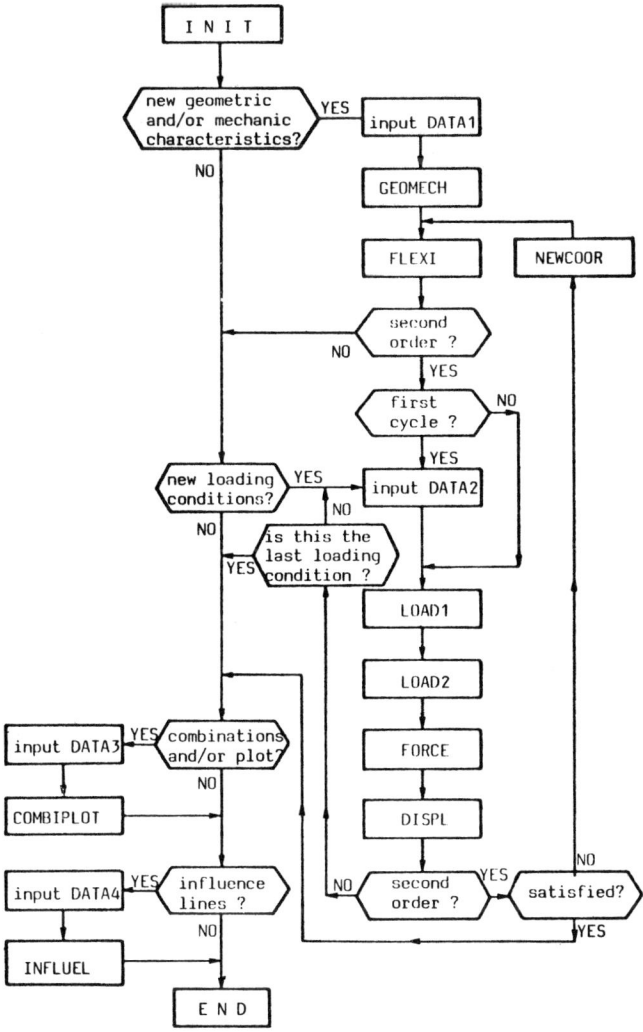

Fig. 1. Block flow diagram of code AFAG.

considered (N.B. 1 kgf = 9.81 N; 1 t = 1000 kg) stability analysis must be performed taking geometrical nonlinearities into account. The analysis was performed automatically by means of iterations; the bending moment diagrams reveal the difference between a Ist order analysis and a 2nd order analysis using AFAG.

This is an example of application to slender structures made up of compressed, deformable elements requiring analysis of safety against instability. Instabiliting forces may derive from wind, seism, etc.

TABLE I Modules for AFAG Programme

input DATA1:	Geometrical characteristics: span, rise, elastic modulus E, shear factor, material unit weight, Poisson ratio, thermal expansion coefficient, number of elements, geometry code, constraint code, characteristics of sections A_i, J_i, h_i (i=1...n), cross section variation code.
input DATA2:	External loads, dead load, spread loads, concentrated loads, V, H, M. Impressed deformations: thermal variations, inelastic displacements, distorsions.
input DATA3:	Data on combination features, deformation parameters to be printed or plotted, displacement amplifier for tracing of deformed shape of single and combined load conditions.
input DATA4:	Data on cross sections and force and deformation parameters for which influence lines are required.
INIT	: Initialling of codes. Explanation of programme use.
GEOMECH	: Construction of geometrical and mechanical characteristics of elements.
NEW COOR	: Updating of joint coordinates.
FLEXI	: Construction of flexibility matrix.
LOAD1	: Construction of known terms for external loads. Print zero state forces.
LOAD2	: Construction of known terms for impressed deformations.
FORCE	: System resolution. Calculation of forces and print out.
DISPL	: Calculation of the components of joint displacements and print out.
COMBIPLOT	: Combination of two or more load conditions: print and/or plot of forces and/or deformed shape. Plot of forces and/or deformed shape of single load conditions.
INFLUEL	: Calculation, print out and plot of influence lines for vertical or horizontal travelling load for force parameters of any section.

Figure 2 shows the structural scheme, deformed shapes and bending moment diagrams.

Discretization: the two lower columns were subdivided into 12 elements and the pitch-beams into 8 elements.

The elements are beam elements liable to deformation by normal and shear forces; there are 62 elements and 61 joints in all, but the degree of indetermination (static indetermination) is always 3.

The analysis was performed with AFAG on HP 9831A, HP 9871A printer, and HP 9872A plotter.

The input and output are extremely articulated and exhaustive to give a complete description of the analysis; they are not reported for reason of space.

Concrete Circular Bridge Arch

Figure 3 shows the analysis of internal forces and deformations of a circular arch fixed at the bases and with a hinge at the apex, submitted to uniformly distributed loads. The geometrical dimensions are shown. The arch is made of concrete (E = 2.10^7 kN/m^2; material unit weight = 25 kN/m^3; Poisson ratio = 0.2), cross sections are constant along the axis (area A = 0.32 m^2, inertia moment I =

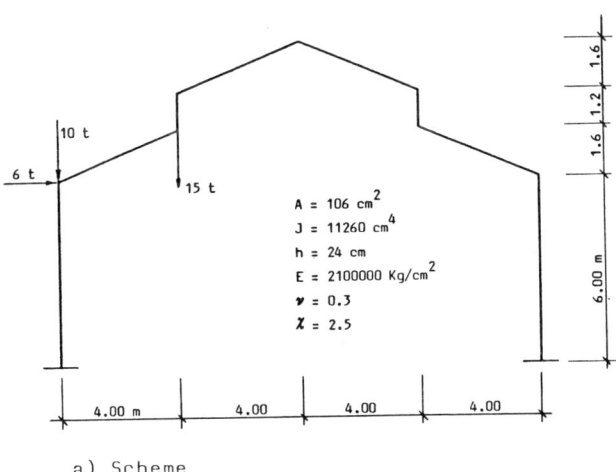

a) Scheme

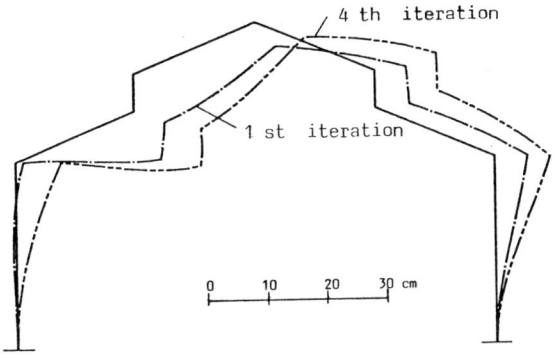

b) Deformed shapes

Fig. 2. a) and b) Steel portal frame.

0.1184 m⁴, height h = 1.20 m).

The example concerns civil engineering, and industrial buildings and bridges in particular.

The arch considered was subdivided into 100 elements, liable to deformation by bending moment, axial and shear forces.

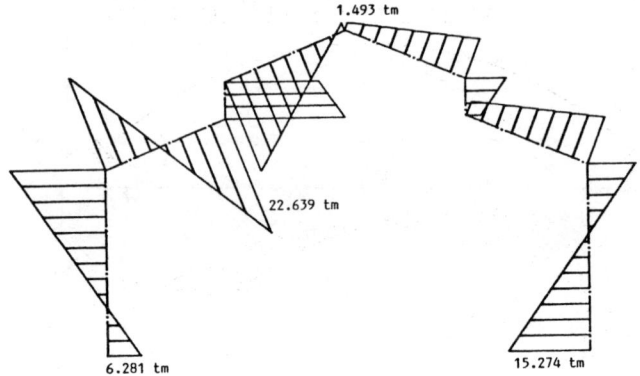

c) First order bending moment diagram

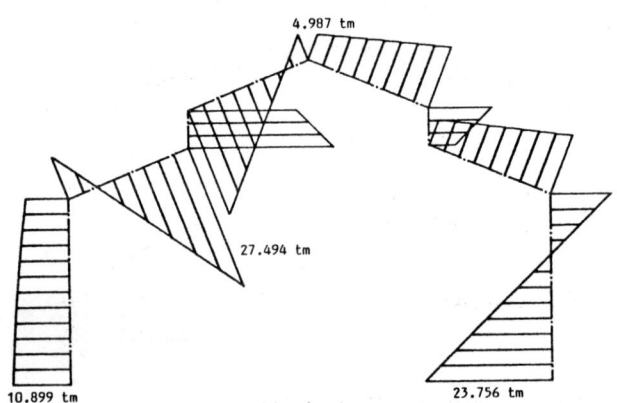

d) Second order bending moment diagram

Fig. 2. c) and d) Steel portal frame.

There are 99 joints and 3 unknowns.

The calculation equipment is as above. The analysis was performed with AFAG on HP 983aA desktop computer, HP 9871A printer, and HP 9872A plotter.

The input and output are not reported here for reason of space.

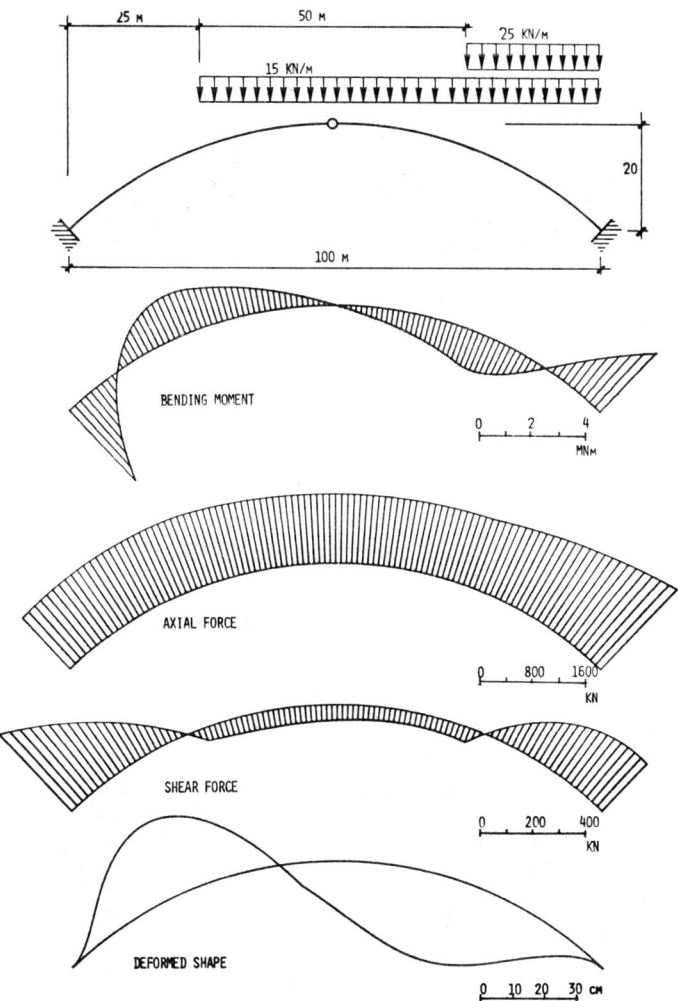

Fig. 3. Circular bridge arch.

Parabolic Bridge Arch. Influence Lines

Figure 4 shows a parabolic arch fixed at the bases.

The analysis presented in this example concerns the drawing of influence lines of the most important parameters of force.

The structure is subdivided into 100 elements liable to deformation by bending moment, axial and shear forces.

The example relates to civil engineering, in particular to the design and analysis of bridges.

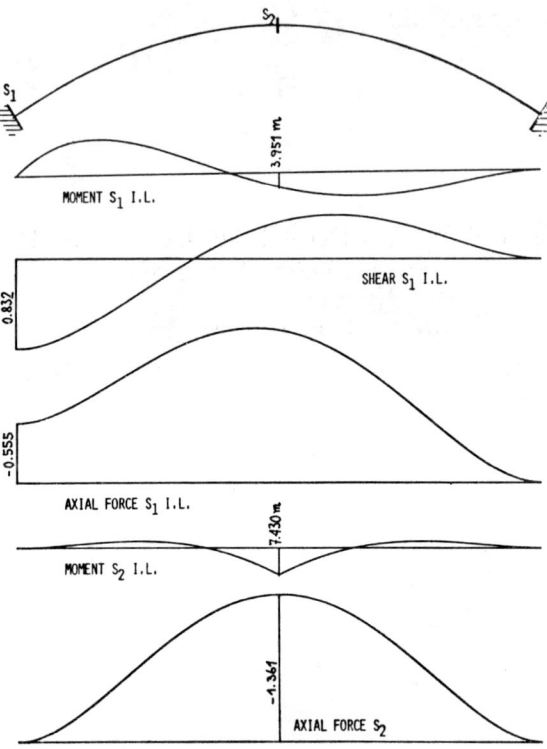

Fig. 4. Parabolic bridge arch.

The cross sections are constant.

AQUADYN: A LINEAR MODELLING TOOL IN HYDRODYNAMICS

J. P. Le Fèvre

CISI Petrole Services, B.P. 83, 92500 Rueil Malmaison, France

ABSTRACT

The AQUADYN system is devoted primarily to the solution of the fully linearized wave scattering/radiation problem for large floating or immersed structures, either isolated or consisting of multibodies interacting together through the waves they radiate towards one another. As an option, AQUADYN will also solve the linearized equations of motion of the structures in the frequency domain, providing the transfer functions between wave excitation and small amplitude motion.

REDUCTION OF THE HYDRODYNAMICS PROBLEM TO A LINEAR INTEGRAL EQUATION

In order to linearize and uncouple the general structural hydrodynamics of floating or immersed bodies (close to the free water surface) the following assumptions are made in AQUADYN:

(a) the structures are rigid.

(b) the fluid is incompressible and has no viscosity.

(c) the fluid motion is a irrational and therefore the velocity field in the fluid is the gradient of a potential Φ satisfying Laplace's equation.

(d) the motion of the structures consisting of three translations and three rotations is steady state sinusoidal with small amplitude about a fixed average position (no forward speed).

(e) the sea has constant, finite or infinite depth; the sea is infinite horizontally.

(f) the exciting gravity waves are steady state sinusoidal AIRY waves with small amplitude.

(g) the boundary conditions on the structures (continuity of the normal component of velocity at the fluid structure interface) are applied to the reference, motionless, configuration thereof instead of the instantaneous displaced configuration; this assumption is acceptable only if motion amplitudes are small

(h) the free surface boundary condition at the water/air interface is linearized and is further applied to the still water plane instead of the actual instantaneous free surface; this assumption is acceptable provided the exciting, scattered and radiated wave amplitudes are small. When all these assumptions hold, it can be shown that the problem reduces to that of finding the potential function $\Phi(x,y,z,t)$ where x,y,z denotes a fixed coordinate system, satisfying the linear integral equation:

$$\frac{\Phi(\mathbf{x})}{2} = \int_{Sh} \left[\Phi(\zeta) \frac{\partial G(\mathbf{x},\zeta)}{\partial n_\zeta} - f(\zeta) G(\mathbf{x},\zeta) \right] dS$$

Sh is the wetted surface of the hull; $\mathbf{x}$ and ζ are the coordinates of a couple of points belonging to the hull; $f(\zeta)$ is the known distribution of normal velocity on the hull; $G(\mathbf{x},\zeta)$ is the (singular) Green function of the problem, which is the potential generated at point $\mathbf{x}$ by a pulsating source at point ζ and obeying the free surface boundary conditions at the air water interface and the zero normal velocity condition on the sea bottom together with an appropriate radiation boundary condition at infinity.

Wave Scattering and Wave Radiation

Classically the total potential Φ is separated into three terms as follows:

$$\Phi = \Phi_I + \Phi_S + \Phi_R$$

where

Φ_I is the incident potential of the undisturbed AIRY waves

Φ_S is the diffraction potential of the waves scattered by the fixed motionless structure(s)

Φ_R is the radiation potential generated by the steady state sinusoidal motion of the structure in a flat sea.

Linearized Structural Dynamics

If further linear structural dynamics calculations must be done, they will involve:

(a) the specification of the mass matrix (Mi), hydrostatic restoring force/moment matrix (Hi), mooring/anchorage linearized stiffness matrix [Ki], mooring/anchorage viscous damping matrix [Ci] for each body (i) and (possibly) elastic and damping coupling matrices [Kij], [Cij] for all couples (i,j) of linked structures.

(b) the calculation of the added mass [MAi] and radiation damping [CRi] matrices associated with the radiation potential ϕ_R.

(c) the calculation of the wave exciting forces

$$\{F_{Ei}\} \text{ and moments } \{M_{Ei}\}$$

due to the exciting potential ϕ_E acting upon each structure (i).

(d) the assembly of the linear 6n*6n (complex) impedance matrix at constant circular frequency ω.

$$[Z(\omega)] = -\omega^2 [M+M_A] + i\omega [C+C_R] + [K+H]$$

(e) the assembly of the 6n (complex) generalized force/moment amplitude vector

$$\{F(\omega)\}^T = \{F_{E1}^T\ M_{E1}^T\ F_{E2}^T\ M_{E2}^T \ldots F_{En}^T\ M_{En}^T\}$$

(f) the solution of the complex linear system $[Z]\{Q\} = \{F\}$ for the generalized 6n (complex) displacement/rotation amplitude vector

$$\{Q(\omega)\}^T = \{T_1^T\ \Omega_1^T\ T_2^T\ \Omega_2^T \ldots T_n^T\ \Omega_n^T\}$$

where superscript T denotes vector transposition and N the number of independent bodies.

FIELD OF APPLICATION

AQUADYN is devoted to the solution of the wave scattering/radiation problem for large floating or immersed structures, either isolated or consisting of multibodies interacting together. Following results are provided:

- geometrical and hydrostatic properties
- added masses and damping coefficients, exciting forces
- steady state sinusoidal motions and resonance
- hydrodynamic pressure on the hull
- drift forces

HYDRODYNAMIC DISCRETIZATION IN AQUADYN

In order to numerically solve the integral equation reported in section 2.1, the wetted surface of *the hull Sh is discretized into a finite number of flat panels* also known as boundary elements. The panels can be either quadrilateral (the preferred option) or triangular.

Further, *the single layer potential $f(\zeta)$ and double layer potential $\Phi(\zeta)$ are assumed constant over each individual panel*.

Let then m denote the total number of panels of the discrete model, and Φ_j the value of the potential for each panel (j); approximating the surface integral by a mere sum of contributions of the m panels, and expressing the integral equation successively at the centroid xi of each panel (i) leads to the system of linear algebraic equations with complex coefficients (since the Green function is complex valued)

$$\tfrac{1}{2}\Phi_i - \sum_1^m \Phi_j \int_{S_j} \frac{\partial G(x_i,\zeta_j)}{\partial n_j}\, dS = -\sum_1^m \int_{S_j} f(\zeta_j) G(x_i,\zeta_j)\, dS$$

where Sj denotes the surface of panel (j). The size of this system being m2 (complex), the fully populated system matrix storage requires 4m2 words of computer memory.

In AQUADYN the influence coefficients

$$C_{ij} = \tfrac{1}{2}\delta_{ij} - \int_{S_j} \frac{\partial G(x_i,\zeta_j)}{\delta n_j}\, dS$$

and right hand side terms

$$\int_{Sj} f(\zeta_j) G(x_i, \zeta_j) dS$$

are integrated analytically (exact integration) after, however, the exact Green function has been approximated in terms of the complex exponential integral, a process which itself involves numerical integration over a single, auxiliary, angular variable.

Internal Architecture of the AQUADYN System

The AQUADYN suite consists of a set of FORTRAN 77 subroutines.

Modules communicate together via sequential FORTRAN work files some of which are formatted, the other ones binary. Linking of the modules is done, using the programmable job control language provided by the COS Operating System. This involves a two level nested looping on wave periods and wave directions of propagation.

The function of each module is given below:

MODG Complete input processing; data decoding and check for consistency and syntaxic correctness; hydrostatic coefficients calculations; geometric properties of hull panels.

COPG Hydrodynamic influence coefficients

MUNG Unit motion radiation potential evaluation

CMAG Added mass and radiation damping matrices for structural dynamics evaluation

PREG Scattered potential evaluation exciting forces

MOUG Linear structural dynamics, hydrodynamic pressures

Within this suite of modules, the most CPU intensive step corresponds to COPG; the largest central memory requirement occurs in the MUNG step.

External Architecture of the AQUADYN System

The AQUADYN system architecture consists of a modular main suite of programs and a number of preprocessors. The preprocessors and main suite communicate via formatted sequential FORTRAN files.

USING AQUADYN SYSTEM

AQUADYN is accessed via interactive procedures allowing unexperienced users to run the program quite easily. The user communicates with the system via performatted screens that give him access to several options: data file edition, data check, full execution, restarts, etc.

No particular knowledge of control language is needed to run AQUADYN.

Input data consists of free format cards with non positional keywords. Extended data check is performed before calculation. Internal mesh generation facilities are available. External meshfile provided by a mesh generator can be used.

Geometrical symmetries are taken into account. Safeguard against time unit is implemented. An output file for processing is provided. There are no imbedded model size limitations in AQUADYN. Most array dimensions are parametrized such that a particular version of the program can easily be generated that exactly suits the array size requirements for each individual run, at the expense of a partial program recompilation at each run.

HARDWARE COMPATIBILITIES

The total computer memory requirement depends on the discretization used for a calculation and can be adapted. On the CISI Network, AQUADYN is available in CRAY XMP version (compiler CFT 1.11, operating system COS 1.12). This version makes full advantage of array processing (acceleration factor 10). A VAX version (compiler FORTRAN 77, operating VMS 3.6) is also available. An IBM version and an HP9000 version will be also run soon.

EXAMPLE

Hydrodynamic analysis of a concrete drilling semi-submersible (Courtesy of C. G. Doris, Fig. 1).

AQUADYN hydrodynamic pressure plot

Instantaneous values at t/T = 0.30

Fig. 1.

Description   the body to study consists of two pontoons, about 140 m long
                          and 34 m large, two horizontal bracings and six columns. The
                          displacement is 17,000 m^3 and the draught 40 m.

Application area: offshore

Problem   diffraction radiation analysis: calculation of the motions,
                          hydrodynamic forces, drift forces, pressures on the hull.

Modelling   because of the symmetries, only a quarter of the structure was
                          discretized with 100 elements.

Calculations   using AQUADYN, calculations were carried out on the CRAY XMP.
                          For one period, computation time was about 30 seconds.

ASE: ANALYSIS OF SPACE STRUCTURES

C. Katz-Fink-Stieda

Römerweg 1, D-8138 Andechs-Erling, Federal Republic of Germany

PROGRAM DESCRIPTION

Program ASE stands for the static or dynamic analysis of spatial structures like shells or solids. Special versions for slabs or plane plates are available.

The program is based on a displacement model. The structure has to be discretized in nodes, which can have rigid or elastic supports, and beam truss, flat shell, or solid elements. Rigid elements are included.

The shell element is an orthotropic 4 node element with a Mindlin bending theory and a classical plane stress approach. The elements can have elastic support by subgrade modulus.

A special pile element with geometric and material nonlinear effects, anisotropic bedding and loading between the nodes is implemented.

Loading includes nodal forces and moments as well as surface and volume loading of the elements. Temperature, earth or water pressure and displacement loading are available.

Dynamic analysis comprehends eigenvalue analysis and modal response.

The modular package consists of several programs which are connected by a database.

AQUA	Generation of cross sections for beam elements
GENF	Mesh generation
OPTIMA	Optimization of mesh numbering
ASE	Static or dynamic analysis
MAXIMA	Superposition of load cases
AQUB	Design of cross sections
BEMESS	Design of shell elements (reinforced concrete)
GRAF	Graphical display of results
SEPP	Special version for plane slabs
ASS	Special version for plane strain and rotational symmetry

HANDLING

The program normally will run as batch job. The input is free format and has generation facilities. The input files can be created via a standard editor program or via a guided input with full screen options. Mesh generation with super elements is supported.

FIELD OF APPLICATION

The program ASE has its main applications in the analysis of plate and shell structures in civil engineering.

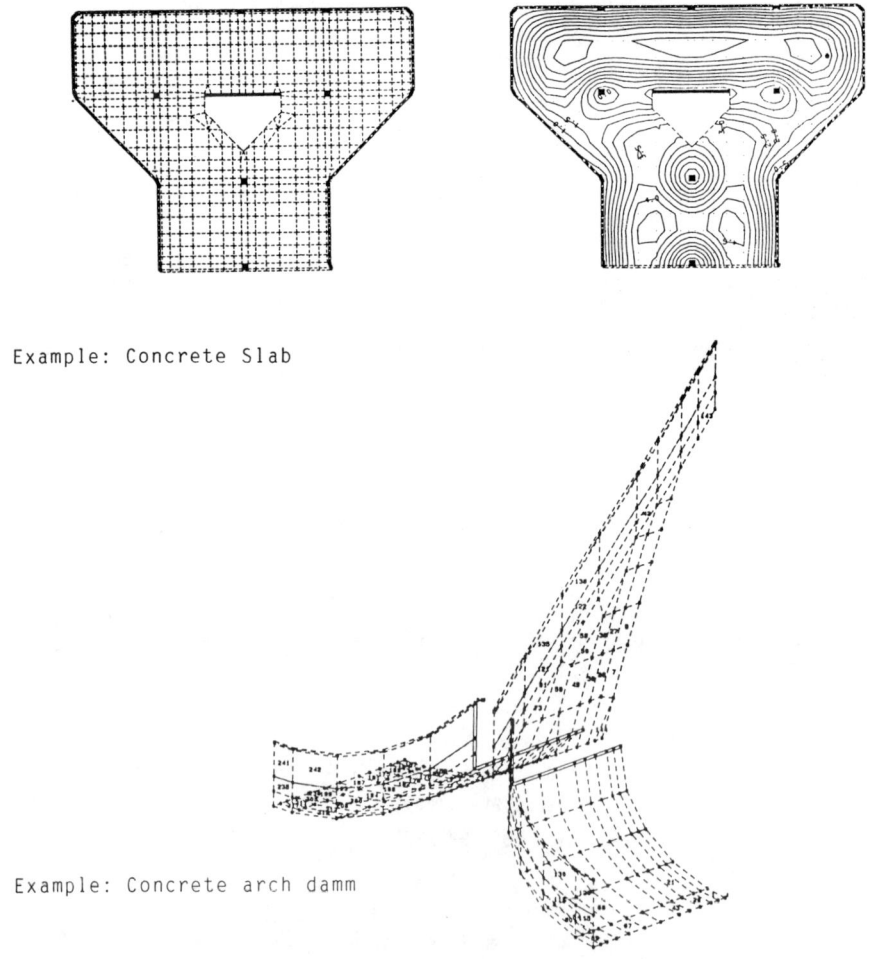

Example: Concrete Slab

Example: Concrete arch damm

HARDWARE REQUIREMENTS

The program is written in FORTRAN-77 and will run on nearly any computer providing that language. Ready to run versions are sold for CP/M-80 and MS-DOS micro-Computers. 2 MByte mass-storage capacity (hard disk) is required. Graphics use the GKS standard functions.

BEMFFT: A PROGRAM FOR THE NUMERICAL SOLUTION OF ONE-DIMENSIONAL BOUNDARY INTEGRAL EQUATIONS

U. Lamp

T. H. Darmstadt Fb4, Schlossgartenstr. 7, 6100 Darmstadt, Federal Republic of Germany

ABSTRACT

Our program solves one-dimensional boundary integral equations on closed smooth curves. Trigonomial polynoms serve as trial and test functions in a Galerkin procedure. For the numerical integration we use a modified fast Fourier transform. The program is applied to an exterior scattering problem with given boundary tractions. It can easily be extended to other boundary integral equations. The numerical results are very accurate and are in good agreement with the theoretical error estimates which have been obtained by U. Lamp, K. T. Schleicher and W. L. Wendland.[1]

THEORETICAL BACKGROUND

Scattering at a Hard Obstacle

The classical scattering problem with a hard scatterer is modelled by the exterior Neumann problem for the Helmholtz equation with the Sommerfeld radiation condition, i.e.

$$\Delta U + \kappa^2 U = 0 \quad \text{in} \quad \mathbb{R}^2 \setminus \bar{\Omega}$$

$$\left.\frac{\partial U}{\partial \nu}\right|_\Gamma = \phi ,$$

$$U(x,y) = \sigma(\frac{1}{\sqrt{r}}) \text{ and } \frac{\partial U}{\partial r} - i\kappa U = O(\frac{1}{\sqrt{r}}) \text{ for } r = (x^2+y^2)^{1/2} \to \infty .$$

Here we use the direct method which yields a *Fredholm integral equation of the second kind* on the boundary Γ.

The Clamped Plate

The solution U of the boundary value problems for the clamped plate,

$$\Delta^2 U = 0 \text{ in } \Omega , \quad U|_\Gamma = \phi_1 , \quad \left.\frac{\partial U}{\partial \nu}\right|_\Gamma = \phi_2$$

is sought in the form of the potential

$$U(x,y) = \int_\Gamma (\frac{\partial F}{\partial \xi} g_1(\xi,\eta) + \frac{\partial F}{\partial \eta} g_2(\xi,\eta))ds - \vec{a} \cdot \binom{x}{y} + \gamma$$

with $F(x,y;\xi,\mu) = -1/2 \, r^2 \log r$ with $r = \sqrt{(x-\xi)^2 + (y-\eta)^2}$. The densities $\vec{g} = (g_1,g_2)^T$ are determined by solving a system of *Fredholm integral equations of the first kind* with logarithmic kernel.

The Traction Problem of Elastostatics

The classical traction problem of plane elastostatics is governed by the Navier equations

$$(\mu\Delta + (\lambda+\mu) \text{ grad div})\vec{u} = 0 \text{ in } \Omega$$

for the displacement field $\vec{u}(x)$ and the boundary condition

$$T[\vec{u}]|_\Gamma = \lambda\vec{\nu} \text{ div } \vec{u} + 2\mu \frac{\partial u}{\partial \nu} + \mu \binom{\nu_2}{-\nu_1} (u_{2x}-u_{1y})|_\Gamma = \vec{f} .$$

The Betti formula for the displacement

$$\vec{u}(x,y) = \int_\Gamma (T_\xi G(x,y); \xi,\eta))^T \vec{u}(\xi,\eta) - G(x,y; \xi,\eta)\vec{f}(\xi,\eta))ds_\zeta$$

$$\text{for } (x,y) \in \Omega$$

- where $G(x,y)$ is the Kelvin fundamental solution - yields on the boundary curve a system of *Cauchy singular integral equations*. In all three cases the boundary integral equations take the form

$$Au + B\omega = f ,$$

$$\Lambda u = \beta ,$$

where u, ω denote a vector valued unknown function on the boundary Γ and an unknown constant vector, respectively, f and β are a given vector valud function and a constant vector, respectively. A is a matrix of operators, B a matrix of functions and Λ a vector of bounded linear functionals.

Our method is based on Fourier expansions which we write as

$$f(t) = \sum_{k \in \mathbb{Z}} \hat{f}(k)e^{2\pi i k t}, \text{ where } \hat{f}(k) = \int_0^1 e^{-2\pi i k t} f(t) dt .$$

Using the Galerkin method we derive the following quadratic system of linear equations

$$\sum_{j=-N}^{N} \hat{u}_N(j) \, (Ae^{2\pi i j \cdot}, e^{2\pi i k \cdot})_o + (B,e^{2\pi i k \cdot})_o \omega_N = \hat{f}(k)$$

$$\text{for } k = -N,\ldots,N \text{ and}$$

$$\sum_{j=-N}^{N} \hat{u}_N(j) \, \Lambda e^{2\pi i j \cdot} = \beta .$$

(1)

BEMFFT: A Numerical Solution of One-Dimensional Boundary Integral Equations 21

The numerical integration in (1) is performed for the singular kernels explicitly and for the smooth remainders via a modified fast Fourier transform.

As was shown in 1., the methods converge of optimal order or even exponentially.

FIELD OF APPLICATION AND PROGRAM DESCRIPTION

The program treats 2D problems of linear elastostatic and acoustic theory with the boundary element method. Trigonometric polynomials serve as trial and test functions.

As input data a parametrization of the boundary curve Γ and the given boundary data are required as subroutines: $x(t)$, $y(t)$, $f(t)$ $0 \le t \le 1$.

The program is coded in FORTRAN IV and installed under VMS 3.7 on a VAX 11/780 at the Technische Hochschule Darmstadt, Germany. The program has a simple linear block structure with easily exchangeable subroutines. The program is available on a magnetic tape.

EXAMPLES OF TESTS

As a test example for the scattering problem we chose

$$\phi = \sin 2\pi t, \quad \Gamma : x = \cos 2\pi t, \quad y = \sin 2\pi t.$$

The exact solution is given by

$$U(x,y) = \frac{H_1^{(1)}(\kappa \tilde{r})}{\kappa \cdot H_1^{(1)'}(\kappa)} \cdot \frac{y}{\tilde{r}} \; ; \; \tilde{r} = \sqrt{x^2+y^2}$$

TABLE 1   Errors in Example 1

N	error at the point (2,3)
1	$6.98 \cdot 10^{-2}$
3	$4.58 \cdot 10^{-3}$
7	$5.35 \cdot 10^{-4}$
15	$6.58 \cdot 10^{-4}$
31	$8.20 \cdot 10^{-5}$

At a test example for the clamped plate we chose

$$\vec{f} = (2x, 2)^T, \quad \vec{b} = \vec{0},$$

$$\Gamma : x(t) = \cos 2\pi t, \quad y(t) = \sin(2\pi t) \cdot (2 - 1.5 \sin 2\pi t).$$

Here the exact solution is given by

$$U(x,y) = x^2 + 2y.$$

TABLE 2   Errors in Example 2

N	error at the point (-0.5, -2)
1	$3.15 \cdot 10^{-1}$
2	$2.16 \cdot 10^{-2}$
3	$3.51 \cdot 10^{-4}$
4	$4.65 \cdot 10^{-4}$
5	$1.42 \cdot 10^{-4}$
6	$1.48 \cdot 10^{-5}$
7	$9.54 \cdot 10^{-7}$

As a test example for the traction problem of elastostatics we chose the displacement field

$$\vec{u}(x,y) = \left| \begin{array}{c} -\log((x-2)^2 + y^2)^{1/2} + \frac{\lambda+\mu}{\lambda+3\mu} \frac{x^2}{((x-2)^2+y^2)} \\ \frac{\lambda+\mu}{\lambda+3\mu} \frac{(x-2)y}{((x-2)^2+y^2)} \end{array} \right|$$

in the unit circle, in particular $\lambda = \mu = 0.4$. For the boundary conditions we computed $T[\vec{u}]\big|_\Gamma$ explicitly and chose

$$\vec{b} = \left| \begin{array}{c} -1.6062786067 \\ 0 \end{array} \right| .$$

TABLE 3 Errors in Example 3

N	$\|u_1-\tilde{u}_{1N}\|_{L^2}$	$\|u_2-\tilde{u}_{2N}\|_{L^2}$ on Γ
1	$1.32 \cdot 10^{-1}$	$3.83 \cdot 10^{-2}$
2	$5.09 \cdot 10^{-2}$	$1.91 \cdot 10^{-2}$
3	$2.16 \cdot 10^{-2}$	$9.57 \cdot 10^{-3}$
4	$9.64 \cdot 10^{-3}$	$4.78 \cdot 10^{-3}$
5	$4.43 \cdot 10^{-3}$	$2.39 \cdot 10^{-3}$
6	$2.07 \cdot 10^{-3}$	$1.20 \cdot 10^{-3}$
7	$9.83 \cdot 10^{-4}$	$5.98 \cdot 10^{-4}$
8	$4.71 \cdot 10^{-4}$	$2.99 \cdot 10^{-4}$
9	$2.27 \cdot 10^{-4}$	$1.50 \cdot 10^{-4}$
10	$1.10 \cdot 10^{-4}$	$7.48 \cdot 10^{-5}$
11	$5.36 \cdot 10^{-5}$	$3.74 \cdot 10^{-5}$
12	$2.62 \cdot 10^{-5}$	$1.87 \cdot 10^{-5}$
13	$1.28 \cdot 10^{-5}$	$9.34 \cdot 10^{-6}$
14	$6.28 \cdot 10^{-6}$	$3.09 \cdot 10^{-6}$
15	$3.36 \cdot 10^{-6}$	$2.12 \cdot 10^{-6}$

REFERENCE

1. Lamp, U., K.-T. Schleicher and W. L. Wendland. The fast Fourier transform and the numerical solution of one-dimensional boundary integral equations. *Numerische Mathematik*, 47, p. 15-38 (1985).

BOSOR4—PROGRAM FOR STRESS, STABILITY AND VIBRATION OF COMPLEX, BRANCHED SHELLS OF REVOLUTION

D. Bushnell

Lockheed Applied Mechanics Laboratory, Department 93-30, Building 255, 3251 Hanover Street, Palo Alto, California 94304, USA

ABSTRACT

BOSOR4 performs stress, stability, and modal vibration analysis of complex, segmented, branched shells of revolution made of elastic material. It can be used to analyze prismatic shells and panels. It performs moderately large deflection axisymmetric stress analysis, small deflection nonsymmetric stress analysis, modal vibration with axisymmetric nonlinear prestress included, and buckling analysis with axisymmetric or nonsymmetric prestress. Symmetric and nonsymmetric buckling modes can be found. There is provision for realistic engineering details such as eccentric load paths, internal supports, arbitrary branching conditions, and a library of wall constructions, including layered orthotropic with temperature-dependent material properties. BOSOR4 is divided into three processors. The user provides input data in an interactive mode. Extensive "HELP" prompts and definitions are available, so that looking up variable definitions in a user's manual is seldom necessary. CALCOMP-type plotting routines are called from a processor that produces plot files. Prebuckling state and buckling or vibration modes are plotted. BOSOR4 runs extremely fast.

THEORETICAL BACKGROUND AND PROGRAM OVERVIEW

The BOSOR4 computer program was developed in response to the need for a tool which would help the engineer to design *practical shell structures*. An important class of such shell structures includes segmented, branched, ring-stiffened shells of revolution. Even if the actual structure is not a shell of revolution, it is usually beneficial to use BOSOR4 in the preliminary design and analysis phases of a project because one can easily and very rapidly obtain good estimates of the behavior. The engineer is thereby guided to make appropriate models for use with general purpose programs, for which the investment in labor and computer time are usually far greater than those required to generate models for and obtain results with BOSOR4.

The complex segmented and branched shells of revolution may have various meridional geometry, wall construction, boundary conditions, ring reinforcements, stringer reinforcements, and types of loading, including nonuniform distributed and line loads and nonuniform temperature distributions.

Table 1 lists the characteristics and status of BOSOR4 as of November 1, 1984.

The program is currently in widespread use and is maintained by the developer. Notices of bugs found are distributed to all known users. BOSOR4 has been thoroughly checked out by comparisons with other known solutions, tests, and by extensive use at many different institutions the world over for more than 10 years.

TABLE 1   BOSOR4 at a Glance

KEYWORDS: shells, stress, buckling, vibration, nonlinear, elastic, shells of revolution, rings, branched, composites, discrete

PURPOSE: To perform stress, buckling, and modal vibration analyses of ring-stiffened, branched shells of revolution loaded either axisymmetrically or nonsymmetrically. Complex elastic wall construction permitted. Buckling and/or stress response to harmonic or random base excitation.

DATE: 1972; most recent update 1985

DEVELOPER: Dr. David Bushnell, 93-30/255, Lockheed Applied Mechanics, 3251 Hanover St., Palo Alto, California 94304, (415)858-4037

METHOD: Finite difference energy minimization; Fourier superposition in circumferential variable; Newton method for solution of nonlinear axisymmetric problem; inverse power iteration with spectral shifts for eigenvalue extraction; Lagrange multipliers for constraint conditions; thin shell theory.

RESTRICTIONS: 3000 degrees of freedom (dof) in nonaxisymmetric problems; 2000 dof in axisymmetric prebuckling stress analysis; maximum of 80 Fourier harmonics per case; knockdown factors for imperfections not included; radius/thickness should be greater than about 10. Up to 80 shell segments permitted, each with its own meridional geometry, wall construction, loading, and constraint conditions; linear elastic material.

Language: FORTRAN 77.

DOCUMENTATION: BOSOR4 User's Manual (Ref. 1) and 11 journal articles and a book with numerous examples (References 2 through 13); Interactive input with on-line HELP. Hierarchical interactive "HELP" file is also available. A file called BOSOR4ST.ORY is included with the program system. This file describes how to use BOSOR4 on the VAX and contains information about how to get the plotting capability "up" at the user's facility. BOSOR4ST.ORY also identifies improvements in the VAX-BOSOR4 capability beyond those of earlier versions of BOSOR4.

INPUT: interactive input. Required for input are shell segment geometries, ring geometries, number of mesh points, ranges and increments of circumferential wave numbers, load and temperature distributions, shell wall construction details, and constraint conditions.

OUTPUT: Displacements and stress resultants or extreme fiber stresses, buckling loads, vibration frequencies; list and plots.

HARDWARE: VAX11/750, 11/780, VMS operating system; CALCOMP-type plot subroutine calls. IBM, PRIME versions also available (see below).

SIZE: 1400 blocks required for storage of BOSOR4 absolute elements; 1600 blocks required for source files; 1200 blocks required for relocatables; from 1000-5000 blocks required for I/O for a typical case.

USAGE: About 200 institutions are using BOSOR4. It is currently being used on a daily basis by many of them. (See Table 3 for partial list.)

RUN TIME: Typically 1 to 15 minutes on VAX11/780, VMS operating system.

AVAILABILITY: VAX version available from developer (address above); Price: $1000.00 includes all documentation, magnetic tape with source, relocatables, absolutes, test cases, further documentation. IBM and VAX versions available from Professor Victor I. Weingarten, Structural Research and Analysis Corporation, 1661 Lincoln Blvd., Suite 100, Santa Monica, California, 90404, Tel.(213)452-2158; Prime Version available from Howard Jaeger, Code 244.5, MS 060, Mare Island Naval Shipyard, Vallejo, California 94592, Tel.(707)646-2444 or -3273. VAX version also available from Professor G. D. Galletly, Mechanical Engineering Department, The University of Liverpool, P.O. Box 147, Liverpool L69 3BX, United Kingdom. Prices of BOSOR4 available from others than the developer not known. One time purchase price.

MAINTENANCE: Developer sends out notices of bugs and other news from time to time.

FIELD OF APPLICATION

The BOSOR4 capability is summarized in Table 2.

TABLE 2 BOSOR4 Capability Summary

TYPE OF ANALYSIS:

1. Nonlinear axisymmetric stress analysis (by itself or for generating prebuckling or prestress states for bifurcation buckling or modal vibration analyses)

2. Linear symmetric or nonsymmetric stress analysis (by itself or for generating prebuckling states for bifurcation buckling analysis)

3. Bifurcation buckling with linear symmetric or nonsymmetric prestress analysis, or with nonlinear symmetric prestress analysis

4. Modal vibration with nonlinear axisymmetric prestress analysis

5. Response to harmonic loading (excitation via applied loads)

6. Response to harmonic or random base excitation

SHELL GEOMETRY AND CONSTRAINT CONDITIONS:

1. Multiple-segment, branched shells, each segment with its own wall construction, geometry, loading, constraint conditions, nodal point spacing and distribution (constant nodal point spacing recommended).

2. Especially simple input for cylinders, cones, plates, spherical, ogival, toroidal, ellipsoidal shell segments

3. General meridional shape: point-by-point input (not recommended unless absolutely necessary)

4. Axisymmetric sinusoidal or random imperfections (small amplitude only!)

5. Axial and radial discontinuities in shell meridian (keep them small!)

6. Arbitrary choice of reference surface location relative to the material in the

shell segment wall (near the material, though!)

7. Branched shells

8. Prismatic shells and composite built-up panels[10]

9. juncture full or partial compatibility; sliding

10. constraints to ground at segment ends or within segments

11. Winkler elastic foundation

WALL CONSTRUCTION:

1. Linear elastic material

2. Monocoque constant or variable thickness

3. Layered orthotropic; composite laminate

4. Layered orthotropic with temperature-dependent material

5. Corrugated semisandwich

6. Any of the above wall types reinforced by stringers and/or rings treated in the model as "smeared out"

7. Any of the above wall types further reinforced by rings treated as discrete beam-type structures with torsional-flexural stiffness

8. Arbitrary wall construction supplied via 6 x 6 integrated constitutive law ($C(i,j)$ matrix)

9. Wall properties varying along the shell meridian

LOADING:

1. Axisymmetric or nonsymmetric temperature distributions

2. Axisymmetric or nonsymmetric pressure, traction components

3. Axisymmetric or nonsymmetric line loads at discrete rings

4. Axisymmetric or nonsymmetric imposed displacements at discrete rings

5. Load sets A (eigenvalue parameter) and B (constant)

6. Harmonic (time-varying) applied loads

7. Harmonic or random base excitation

BOSOR4 performs the following analyses:

1. a nonlinear stress analysis for axisymmetric behavior of axisymmetric shell systems (moderately large deflections, linear elastic material);

2. a linear stress analysis for axisymmetric and nonaxisymmetric behavior of axisymmetric shell systems subjected to axisymmetric and nonaxisymmetric loads;

3. an eigenvalue analysis in which the eigenvalues represent buckling loads or vibration frequencies of axisymmetric shell systems subjected to axisymmetric

loads. Eigenvectors correspond to axisymmetric and to nonaxisymmetric buckling
or vibration modes.

4. BOSOR4 has additional branches corresponding to buckling of nonaxisymmetrically
loaded shells of revolution and corresponding to stress and buckling of shells
subjected to harmonic or random base excitation. These capabilities represent
combinations of the first three analysis types.

Figure 1 shows some examples of branched structures which can be handled by
BOSOR4. Figure 1(a) represents part of a multiple-stage rocket treated as a
shell of seven segments; Figure 1(b) represents part of a ring-stiffened
cylindrical shell in which the ring is treated as two shell segments branching
from the cylinder; Figure 1(c) shows the same ring-stiffened cylindrical shell
with the ring treated as "discrete", that is the ring cross section can rotate
and translate but not deform, as it can in the branched shell model shown in Fig.
1(b). Figures 1(d,e,f) represent branched prismatic shell structures, which can
be treated as shells of revolution with very large average circumferential radii
of curvature, as described in reference 10.

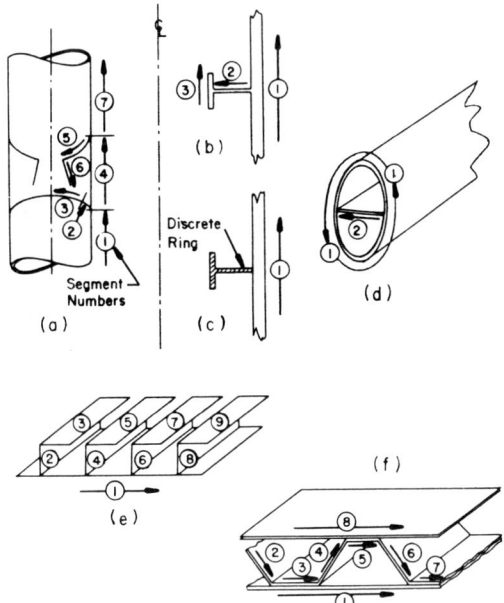

Fig. 1. Examples of branched structures which can be analyzed
with BOSOR4

GOVERNING ASSUMPTIONS IN THE BOSOR4 THEORY

The assumptions upon which BOSOR4 is based are:

1. The wall material is linear elastic.

2. Thin shell theory holds: normals to the undeformed surface remain normal.

3. The structure is axisymmetric, and in modal vibration analysis and nonlinear

stress analysis the loads and prebuckling or prestress deformations are axisymmetric.

4. The axisymmetric prebuckling deflections in the nonlinear theory, while considered finite, are moderate. That is, the square of the meridional rotation can be neglected compared to unity.

5. In the calculation of displacement and stresses in nonsymmetrically loaded shells, linear theory is used. This branch of the program is based on standard small-deflection analysis.

6. A typical cross sectional dimension of a discrete ring stiffener is small compared with the radius of the ring.

7. The cross sections of the discrete rings remain undeformed as the structure deforms, and the rotation about the ring centroid is equal to the rotation of the shell meridian at the attachment point of the ring. (except, of course, if the ring is modelled as a flexible shell branch or branches!).

8. The discrete ring centroids coincide with their shear centers.

9. If meridional stiffeners are present, they are numerous enough to include in the analysis by an averaging or "smearing" of their properties over any parallel circle of the shell structure. Meridional stiffeners can be treated as discrete for prismatic structures, as illustrated in Figs. 1 (d,e,f).

10. In bifurcation buckling analyses of nonaxisymmetrically loaded shells the analysis is carried out in two phases: 1. a linear nonsymmetric stress analysis is first performed; 2. for a user-selected meridian, bifurcation buckling analyses are then conducted. The assumption in this second phase of the analysis is that the prebuckling state of the shell along the user-chosen meridian is the same for all other meridians: it is axisymmetric.

METHOD AND DISCRETIZATION

The BOSOR4 analysis is based on energy minimization with constraint conditions. The total energy of the system includes strain energy of the shell segments and discrete rings, potential energy of the applied line loads and pressures, and kinetic energy of the shell segments and discrete rings. Thermal strains are included. The constraint conditions arise from displacement conditions at the boundaries of the structure, displacement conditions that may be prescribed anywhere within the structure, and compatibility conditions at junctions between shell segments. These constraint conditions are introduced into the energy function by means of Lagrange multipliers.

The components of energy, work, and constraints are initially integro- differential forms. The circumferential dependence is eliminated by separation of variables and the assumption that variation with respect to the circumferential coordinate θ is trigonometric (sin$n\theta$, cos$n\theta$). Displacements and meridional derivatives of displacements are then written in terms of the shell wall reference surface meridional, circumferential, and normal displacement components, u_i, v_i, w_i, respectively, at the finite-difference nodal points. Integration is performed by multiplication of the energy-per-unit-meridional-length by the length L of the "finite-difference element". This unusual kind of finite element is depicted in Fig. 2. It is described more fully in reference 4. The unknowns in the discretized problem are the nodal point degrees of freedom, u_i, v_i, w_i, and Lagrange multipliers, λ_i. Solution of the equation systems corresponding to the various nonlinear and linear simultaneous algebraic equations goes extremely fast on the computer because the average bandwidth of the equations is very small, discretization being only in one coordinate direction. This is the reason that BOSOR4 runs very fast, even on minicomputers.

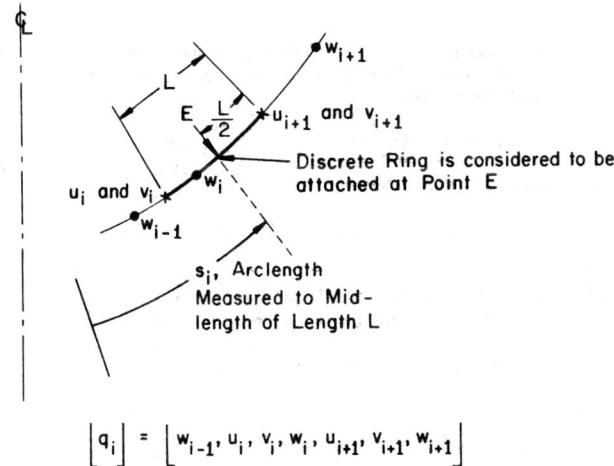

$$[q_i] = [w_{i-1}, u_i, v_i, w_i, u_{i+1}, v_{i+1}, w_{i+1}]$$

Fig. 2. Finite difference discretization: the "finite difference element". The meridional length of the element is L; the single integration point is at the element midlength E; the nodal degrees of freedom involved in the local stiffness, mass, and load-geometric matrices for the ith finite element are identified in the vector q_i.

In the nonlinear axisymmetric stress analysis the energy expression has terms that are linear, quadratic, cubic, and quartic in the dependent variables, u_i and w_i. The cubic and quartic energy terms arise from the rotation-squared terms that appear in the expression for reference surface meridional strain and in the constraint conditions. Energy minimization leads to a set of nonlinear algebraic equations that are solved by the Newton-Raphson method. Stress and moment resultants are calculated in a straightforward manner from the meshpoint displacement components through the constitutive equations and the kinematic relations.

The results from the nonlinear axisymmetric or linear nonsymmetric stress analysis are used in the eigenvalue analyses for buckling and vibration. The "prebuckling" or "prestress" meridional and circumferential stress resultants and the meridional rotation appear as known variable coefficients in the energy expressions that govern bifurcation buckling and modal vibration. These expressions are homogeneous quadratic forms. The values (eigenvalues) of a parameter (load or frequency) that render the quadratic forms stationary with respect to infinitesimal variations of the dependent variables represent buckling loads or natural frequencies. These eigenvalues are calculated from a set of linear homogeneous equations.

Details of the analysis are presented in references 3, 7, 11, 12. The user's manual

USER-FRIENDLY FEATURES OF BOSOR4

The following features make the latest version of BOSOR4 very easy to use:

1. *HIERARCHICAL INTERACTIVE "HELP" UTILITY:* New users of BOSOR4 can learn about what BOSOR4 does, how to use it, how to modify cases, what the various commands do, what is contained in various files generated by BOSOR4 runstreams, etc.

Examples of the use of *HELP* are listed in Appendix 1 at the end of this paper. This appendix contains useful information about BOSOR4, so please read it for information as well as an example of the *HELP* utility.

2. *INTERACTIVE INPUT WITH ON-LINE HELP:* The user provides input data interactively. If the user wants help in connection with any input datum, he or she simply types H(ELP) instead of the datum called for by the interactive prompt. BOSOR4 responds with a paragraph or paragraphs of explanation. Appendix 2 lists a typical BOSOR4 runstream for providing input data for a nonlinear buckling analysis of an externally pressurized spherical cap resting on a frictionless plane. In several instances the user called for H(ELP).

3. *ANNOTATED FILE OF INPUT DATA PRODUCED:* Upon successful completion of the case (actually, successful completion of the preprocessor phase of the case), BOSOR4 produces a file, such as that listed in Appendix 3, that can be modified and used for future similar cases, thereby rendering it unnecessary for the user to answer all the questions interactively again. This file (which in the example of Appendix 3 corresponds to the case of the spherical cap just mentioned) also serves to document the problem.

4. *EDITING UTILITY "MODIFY" FOR MODIFYING CASES:* Page 3 of Appendix 1 describes this feature. Modification is carried out automatically by the BOSOR4 system through use of "key phrases", such as "DISCRETE RING INPUT FOLLOWS" or "LINE LOAD INPUT FOLLOWS" that appear in the annotated file of which Appendix 3 is an example. Small files, such as RINGS.QUE and LINELOADS.QUE, generated interactively when the user gives the command *MODIFY*, are inserted in the annotated input file,

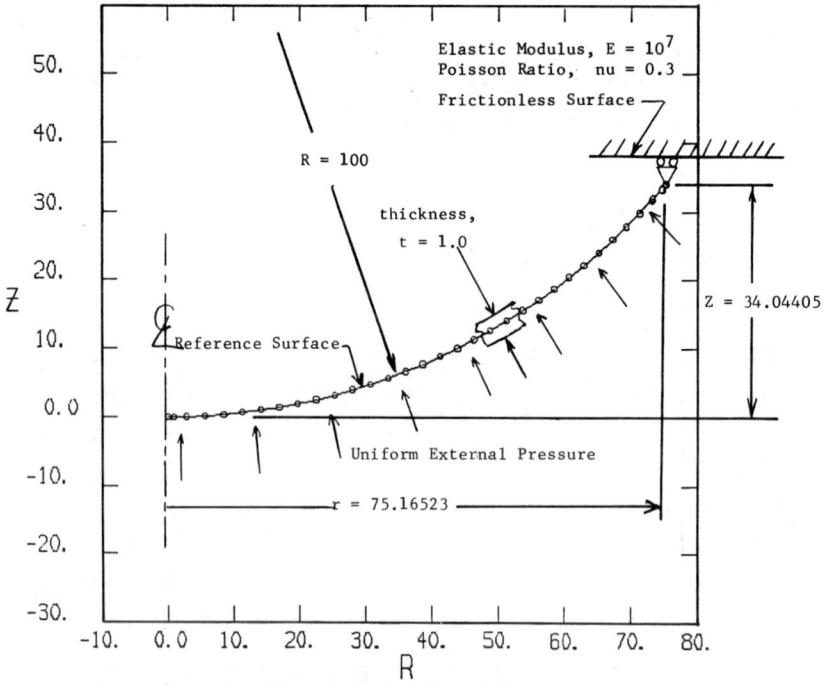

Fig. 3. Sample problem geometry: shallow spherical cap on frictionless plane with uniform external pressure. The nodal points are the locations of the midlengths of the finite difference elements.

replacing earlier entries at these locations. This feature renders it unnecessary to generate new cases from the beginning when there are only a few modifications wanted by the user.

5. *PLOTS:* Plots of the discretized model, prebuckling deflections, and buckling modes are produced via CALCOMP-type subroutine calls. Examples are shown in Figs. 3, 4, and 5, which represent output from the sample case treated as in Appendices 2 and 3. BOSOR4 will also produce "x,y" plots of displacements, stress components and stress and moment resultants, in which "x" is the arc length along the meridian of the shell of revolution.

EXAMPLES OF APPLICATION

DETAILED EXAMPLE CASE: Appendix 2 contains a sample runstream for interactive input of data for a nonlinear buckling analysis of an externally pressurized spherical cap lying on a frictionless plane. Appendix 3 lists the annotated input file produce by BOSOR4 for documentation of the case and for use in future similar cases. The specifics of the case and discretization are shown in Fig. 3. Some results are plotted and listed in Figs. 4 and 5. In order to save space, listed output from this example are not reproduced here. This output is self-explanatory.

In this example boundary conditions and nonlinear geometric effects are very important. A linear bifurcation buckling analysis of the same cap on the same frictionless plane yields a critical pressure p_{cr} = 851 with n_{cr} = 10 circumferential waves. Compare with the nonlinear model, which yields p_{cr} = 248

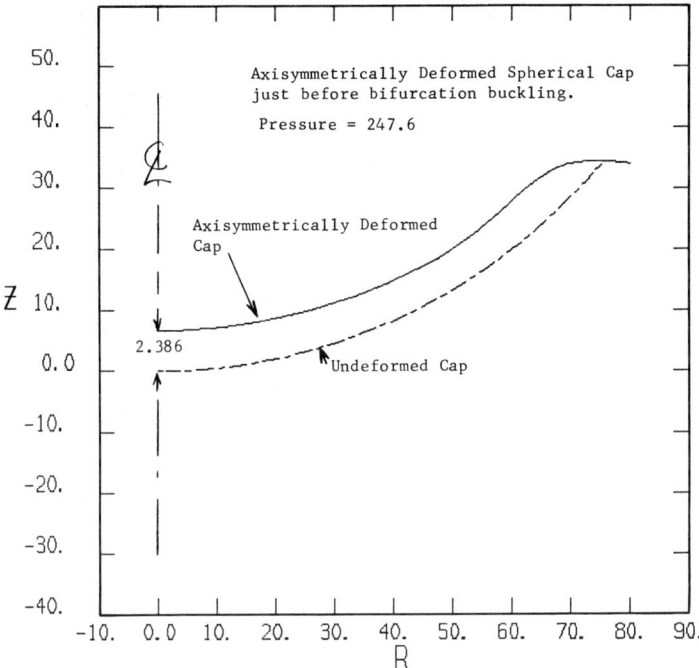

Fig. 4. Axisymmetric prebuckling deformation at an external pressure, p_{cr} = 247.6psi. The normal displacement at the crown is w = 2.386in.

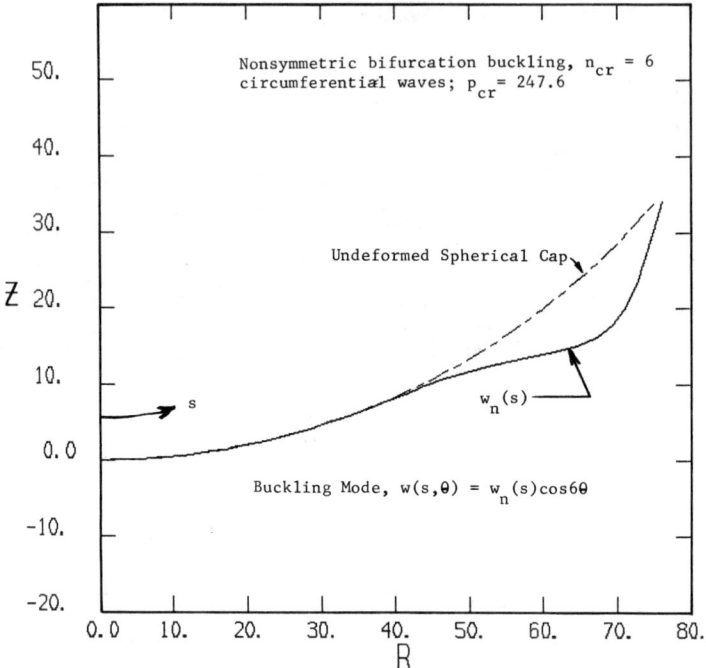

Fig. 5. Bifurcation buckling mode predicted by BOSOR4 for the test problem shown in Fig. 3. Nonlinear geometric prebuckling effects are accounted for in this prediction.

with $n_{cr} = 6$ circumferential waves. The classical buckling pressure of a complete spherical shell with the same radius R, thickness t, and material is $p_{cl} = 1210$.

NONLINEAR STRESS ANALYSIS: Figure 6 shows part of an internally pressurized elliptical tank which has been thickened locally near the equator for welding. The engineering drawings called for an elliptically shaped inner surface with the thickness varying as shown. The maximum stress occurs at the outer fiber at point C because there is considerable local bending there due to the rather sudden change in direction or eccentricity of the load path in the short segment ACB. The nonlinear theory gives lower stresses than does the linear theory because the meridional tension causes the tank to change shape in such a way as to decrease the local excursion of the load path, thereby decreasing the effective bending moment acting at point C. The tank had been built and a linear analysis had been performed. The user of the tank wanted to know if it would withstand a somewhat higher internal pressure than that for which it had originally been designed. The lower stress predicted with nonlinear theory gave this user enough margin of safety to avoid the necessity of redesign.

BUCKLING DUE TO QUASI-STATIC BASE EXCITATION: The United States Nuclear Regulatory Commission is interested in the development of methods for the calculation of buckling loads of large steel reactor containment vessels under various dynamic loads, including those from an earthquake. Figure 7 illustrates the predicted behavior of one such shell in which body forces from $1g$ vertical and $1g$ horizontal ground acceleration components are applied to the structure as if they were static. The buckling load factor $\lambda_{cr} = 21.4$ and critical number of circumferential waves, $n_{cr} = 10$, are computed by BOSOR4 from a model that

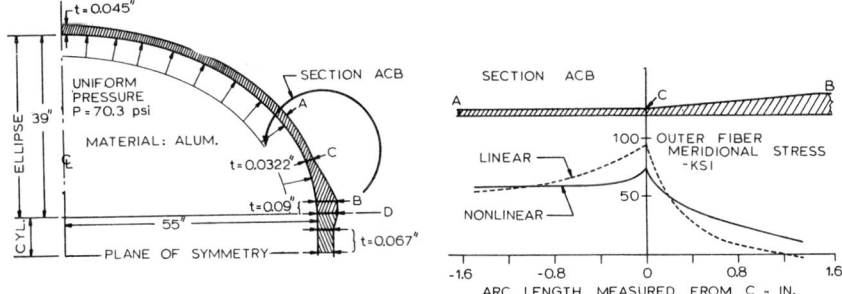

Fig. 6. Linear and nonlinear axisymmetric analyses of an internally pressurized ellipsoidal tank with variable thickness. The stress concentration at Point C is due to load path eccentricity, and the predicted stresses decrease if nonlinear geometric effects are accounted for.

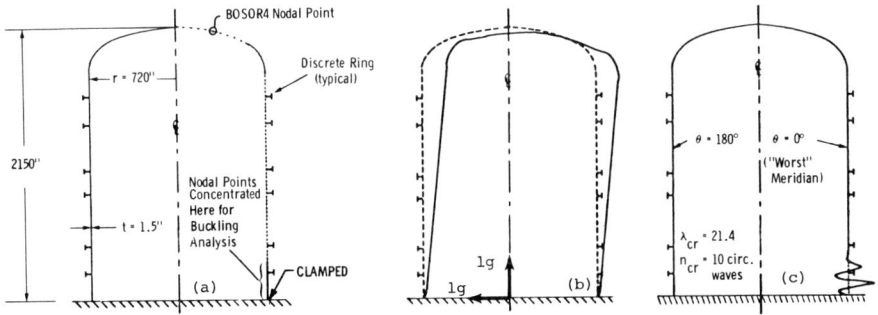

Fig. 7. Buckling of nuclear reactor steel containment vessel due to ground accelerations: (a) BOSOR4 discretized model; (b) prebuckling deformation due to 1g vertical and 1g horizontal ground acceleration components; (c) predicted buckling mode, critical load factor, λ_{cr} = 21.4, and number of circumferential waves, n_{cr} = 10.

accounts for the nonaxisymmetry of loading in the prebuckling phase of the analysis, but treats the membrane stress distribution along the "worst" (most axially compressed) meridian as if it were axisymmetric in the stability phase of the analysis. (The user selects the "worst" meridian. If it is hard to tell which the "worst" meridian is, the user may, through as many "restarts" as he or she wants, find buckling load factors and mode shapes for several meridians.)

UNEXPECTED LOCAL BUCKLING: In Fig. 8 is illustrated a local buckling failure of a large, expensive, semisandwich, corrugated, ring-stiffened payload shroud (a and b) that was subjected to axial compression and bending. The shroud failed unexpectedly during proof testing because of local buckling near a field joint (c and d). In short regions on either side of the field joint, where the external corrugations are cut away as shown in Fig. 8(d), the axial load path is deflected inward from the neutral axis of the cross section of the combined corrugations and skin to the middle surface of the skin and doubler. This local inward excursion of the axial load path creates under axial compression the localized hoop compression that causes local nonsymmetric bifurcation buckling in a mode displayed in Fig. 8(c). Because of the short axial length of the circumferentially compressed region, the critical buckling mode has a rather large number of

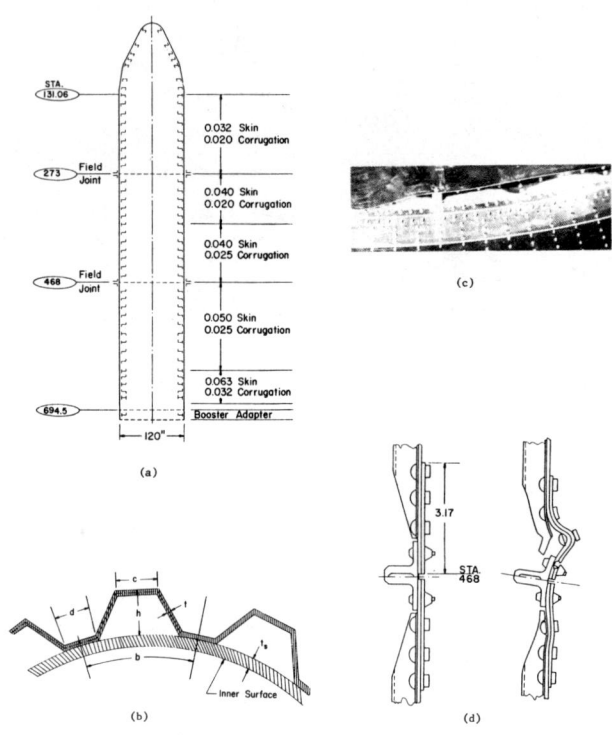

Fig. 8. Local failure of a large payload shroud under axial compression and bending: (a) Typical ring-stiffened rocket payload shroud configuration; (b) Corrugated, semi-sandwich wall; (c) Interior view of portion of complete shroud buckled locally next to field joint at Station 468; (Three waves are visible.) (d) Field joint geometry and buckle configuration. Buckling is caused by the narrow band of circumferential compression arising from the inward excursion of the axial load path near the field joint.

circumferential waves. This mode was obtained by BOSOR4, after the test unfortunately, because those responsible for the test were not aware that this type of local buckling was possible.

BUCKLING UNDER NONSYMMETRIC AERODYNAMIC PRESSURE: Figure 9 shows nonsymmetric pressure loading on the corrugated, ring-stiffened rocket payload shroud displayed in Fig. 9(a,b). The pressure distribution, measured in a wind tunnel, corresponds to a small angle of attack. The payload shroud, attached to a heavy rocket motor stage at its aft end, bends as a beam, as shown in Fig. 9(b). The side under maximum axial compression, the leeward side, buckles between ring stiffeners as shown in Fig. 9(c). Buckling does not occur at the root of the beam because the shell wall is made of thicker gauge material there, as indicated in Fig. 8(a).

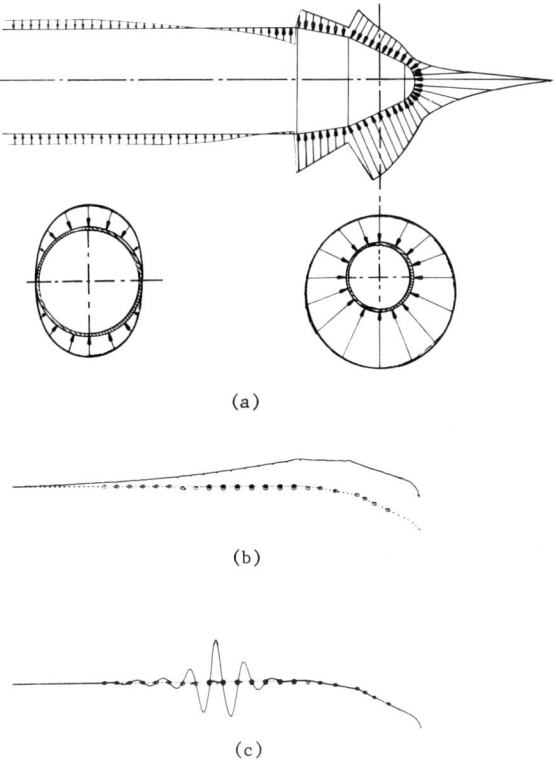

Fig. 9. Buckling mode of nonsymmetrically loaded rocket payload shroud shown in Fig. 9(a): (a) Pressure distribution measured in a wind tunnel test; (b) Prebuckling beam-type deflection; (c) Nonsymmetric buckling mode. Buckling is between the discrete rings and occurs with 13 circumferential waves.

NONLINEAR BEHAVIOR OF INTERNALLY PRESSURIZED FUEL TANK: Figures 10 and 11 pertain to this section. The geometry of the problem is shown in Fig. 10. The tank wall and skirt are divided into segments as indicated. Under small internal pressure the portion of the rocket fuel tank enclosed in the rectangle in Fig. 10 is drawn radially inward, resulting in the development of a narrow band of hoop compression that might lead to bifurcation buckling. At the top of Fig. 11 is shown a bifurcation buckling mode predicted by BOSOR4 with use of linear theory. The modal normal displacement component $w_b(s,\theta)$ varies around the circumference as $\cos 90\theta$.

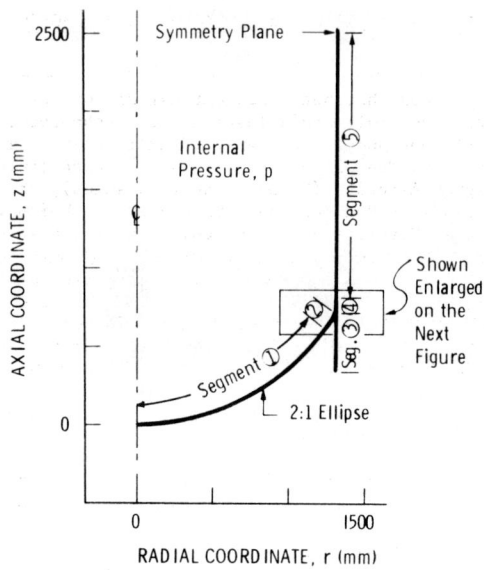

Fig. 10. Bottom part of rocket fuel tank as modelled for input to the BOSOR4 program. Loading is uniform internal pressure. Problem is to determine if the shell buckles in the region enclosed in the small rectangle.

This is a problem for which the use of linear theory in the prebuckling phase of the analysis is inadequate. As the internal pressure is increased the ellipsoidal dome changes shape. The hoop stresses are redistributed and grow more slowly than linearly with pressure, as indicated in the bottom of Fig. 11. As the internal pressure p is increased, the hoop resultant becomes tensile in the region where linear theory predicts bifurcation buckling to occur, and the peak hoop compression initially increases more slowly than predicted by linear theory, eventually reaching a maximum value of about -800 N/mm at a pressure of 1.4 Pa, after which it decreases with further increase in internal pressure. Thus, the prediction by BOSOR4 with nonlinear prebuckling effects included is that bifurcation buckling will not occur at all.

MODAL VIBRATION OF A CRYOGENIC COOLER: BOSOR4 has been in use at Lockheed and elsewhere since 1972. During that time it has been used in several projects, some of them involving rather complex segmented and branched shells of revolution. An example is shown in Fig. 12, which depicts a somewhat idealized model of cryogenic cooler. The axisymmetric structure consists of a series of fiberglass tubes from which are suspended two axisymmetric cryogenic tanks. The object of this study was to determine the natural frequencies of the cooler corresponding to beam-type modes (n = 1 circumferential wave). The segmented and discretized model is shown in Fig. 13 and the first four vibration modes are displayed in Fig. 14. The first two vibration modes and frequencies found by BOSOR4 agreed well with those from tests. The higher two modes were not sought in the test.

WORLDWIDE USE OF BOSOR4

Various versions of BOSOR4 have been in use worldwide since 1972. Table 3 lists many organisations that over the years have used BOSOR4 (and the plastic

BOSOR4: Program for Stress, Stability and Vibration 39

TABLE 3 Applications of BOSOR4 and BOSOR5

AEROSPACE STRUCTURES: BOSOR4 has been used extensively by Jet Propulsion Laboratory, Bell Aerospace, California Institute of Technology, Centre Spatial de Toulouse, France (Ariane booster design), DFVLR in West Germany, Fokker Aircraft Space Division in The Netherlands, Ford Aerospace (antennas), Harris Corp. (antennas), Hughes Aircraft (Venus Pioneer Capsules), Martin Marietta (Space Shuttle Fuel Tanks), SPAR Aerospace in Canada, Technion in Israel, TRW Systems, NASA. Lockheed Missiles and Space Co: Satellite Systems Division has used BOSOR4 extensively over more than 12 years for the design of ring and stringer stiffened shrouds and for the design of frangible joints; Lockheed Missiles Systems Division has used BOSOR4 for buckling predictions of rocket booster equipment sections, nose fairings, ring reinforced composite shells; Lockheed Advanced Systems has used BOSOR4 extensively for the analysis of mirrors for optical systems and for the analysis and design of cryogenic coolers.

AIRCRAFT ENGINES: General electric, Motor-Columbus (Switzerland), Pratt and Whitney, SAAB-SCANIA, Volvo Flygmotor (Sweden).

ROCKET MOTORS: BOSOR4 has been used extensively by Aerojet Liquid Rocket Co., Thiokol, Hercules, UTC, Societe Europeene de Propulsion (France)

AUTOMOBILE WHEEL RIMS: Simpson, Gunpertz and Heger Co.

SUBMARINE PRESSURE HULLS: BOSOR4 and BOSOR5 have been extensively used by U.S. Navy research labs (David W. Taylor NSRDC, NUSC, NOSC); by the French Navy (Service Technique des Constructions et Armes Navales); by the English Navy (Naval Construction Research Establishment); Weidlinger Associates, General Dynamics Electric Boat Division.

CONTAINERS: Continental Can, Alcoa, U.S. Steel (beverage cans); Monsanto, Dow Chemical

CIVIL ENGINEERING STRUCTURES: Aeronautical Research Institute of Sweden (large vessels for storing beer!); Burns and Roe, Chicago Bridge and Iron (large oil tanks); Oxford University (liquid natural gas tanks); Illinois Institute of Technology (corrugated grain storage bins); Envirotech Information Systems, Monsanto, Dow (chemical storage tanks); Princeton University (cooling towers); Universite de Liege, Belgium and Gent University (large water tanks); University of Liverpool (buckling of internally pressurized ellipsoidal and torispherical heads)

NUCLEAR INDUSTRY: Battelle Pacific Northwest Labs, CERN-European Organization for Nuclear Research (Switzerland), Chicago Bridge and Iron (containment vessels), Franklin Institute Research Lab, General Electric, Pittsburgh Des Moines Steel, Stone and Webster Engineering, Union Carbide Corporation Nuclear Division, Westinghouse Electric, Kraftwerk Union (West Germany), U.K. Atomic Energy Authority, U.S. Nuclear Regulatory Commission

OFFSHORE ENGINEERING, SHIPPING: General Dynamics Shipbuilding (spherical LNG tanks); Lloyd's of London, Det Norske Veritas (Norway), Shell KSEPL (The Netherlands)

buckling program BOSOR5). Types of problems solved with the BOSOR programs are indicated.

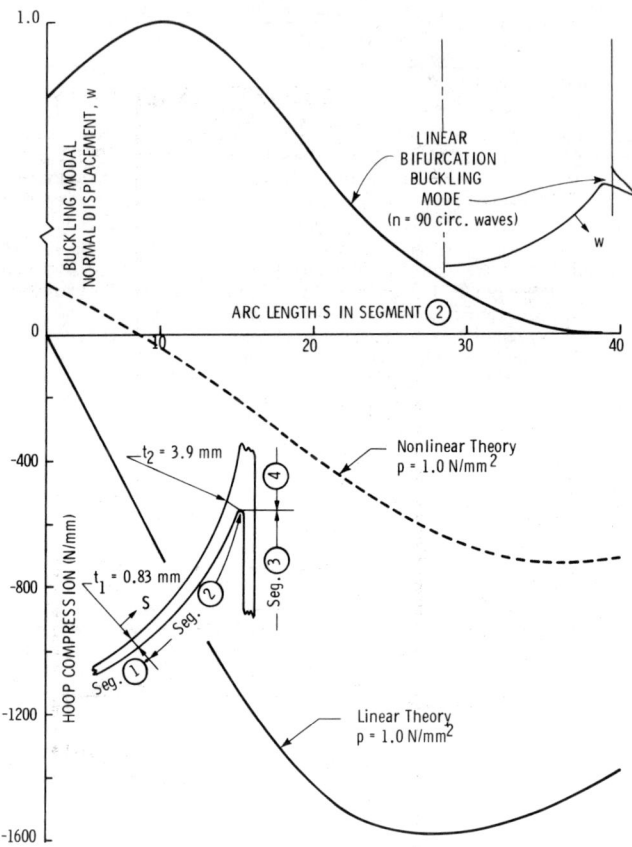

Fig. 11. Nonsymmetric buckling mode (top) and hoop resultants from linear and nonlinear theory (bottom) for the part of the internally pressurized rocket motor enclosed in the small rectangle in the previous figure. Use of linear theory leads to a prediction of buckling, but use of nonlinear theory leads to a prediction that buckling will never occur.

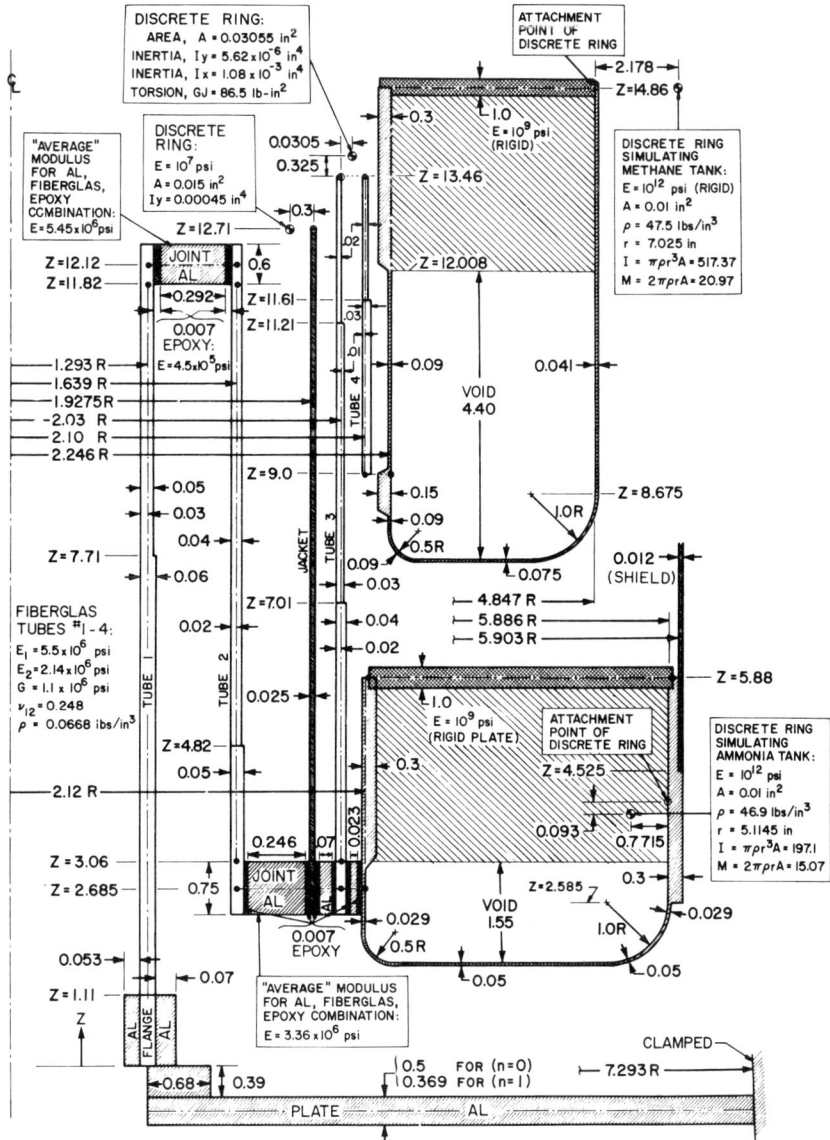

Fig. 12. Cryogenic cooler modelled as a complex, branched shell for analysis by BOSOR4

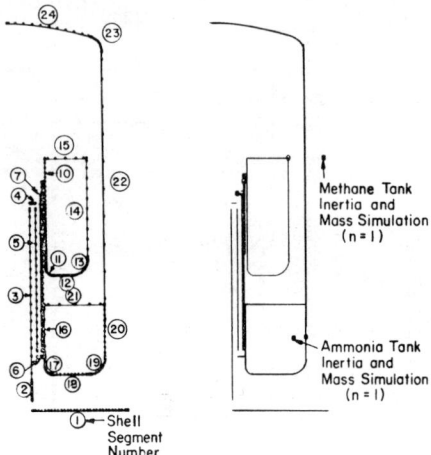

Fig. 13. BOSOR4 segmented, branched, and discretized model. The mass properties of the Methane Tank and Ammonia Tank are simulated by introduction of two discrete rings, as indicated on the right. Half of the cooler is shown.

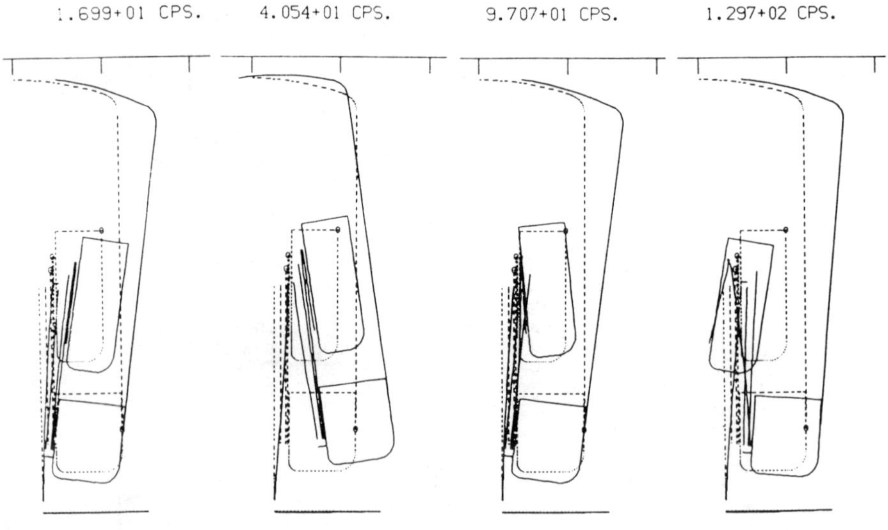

Fig. 14. First four lateral vibration modes predicted with the BOSOR4 model. Lateral vibration corresponds to $n = 1$ circumferential wave. Half of the cooler is shown.

REFERENCES BY D. BUSHNELL PERTANENT TO BOSOR4

1. BOSOR4 - Program for Stress, Buckling and Vibration of Complex Shells of Revolution. Structural Mechanics Software Series, edited by N. Perrone and W. Pilkey, University Press of Virginia, Charlottesville, Va., pp. 11-143, 1977.

2. Buckling of Shells – Pitfall for Designers. *AIAA Journal*, *19*, No. 9, pp. 1183-1226, 1981.

3. Computerized Analysis of Shells – Governing Equations. *Computers and Structures*, *18*, No. 3, pp. 471-536, 1984.

4. Finite Difference Energy Models Versus Finite Element Models: Two Variational Approaches in One Computer Program. *Numerical and Computer Methods in Structural Mechanics*, (edited by Fenves, Perrone, and Schnobrich), Academic Press, New York, pp. 291-336, 1973.

5. Evaluation of Various Analytical Models for Buckling and Vibration of Stiffened Shells. *AIAA Journal*, *11*, No. 9, pp. 1283-1291, 1973.

6. Nonsymmetric Buckling of Cylinders with Axisymmetric Thermal Discontinuities. *AIAA Journal*, *11*, No. 9, pp. 1292-1295, 1973.

7. Stress, Stability, and Vibration of Complex, Branched Shells of Revolution. *Computers and Structures*, *4*, pp. 399-435, 1974.

8. Local and General Buckling of Axially Compressed, Semi-Sandwich, Corrugated, Ring-Stiffened Cylinder. *Journal of Spacecraft and Rockets*, *9*, No. 5, pp. 357-363, May 1972.

9. Stress and Buckling of Nonuniformly Heated Cylindrical and Conical Shells. (with S. Smith), *AIAA Journal*, *9*, No. 12, pp. 2314-2321, 1971.

10. Stress 2004-2013, 1971.

11. Analysis of Ring-Stiffened Shells of Revolution Under Combined Thermal and Mechanical Loading. *AIAA Journal*, *9*, pp. 401-410, 1971.

12. Analysis of Buckling and Vibration of Ring-Stiffened, Segmented Shells of Revolution, *International Journal of Solids and Structures*, *6*, pp. 157-181, Feb. 1970.

13. *Computerized Buckling Analysis of Shells*, Nijhoff, The Netherlands (1985).

APPENDIX 1. EXAMPLE OF USE OF THE HIERARCHICAL INTERACTIVE HELP UTILITY

```
$ HELP4
<BOSOR4>
    Help on various BOSOR4 features can be obtained by typing

        HELP4 TOPIC SUBTOPIC

    in which TOPIC stands for a key word such as COMMAND, DEBUG,
    FILES, OVERVIEW, or PROGRAMS; and SUBTOPIC stands for another
    key word belonging as a subtopic under either COMMAND, DEBUG,
    FILES, OVERVIEW, or PROGRAMS.  For example,

        HELP4 COMMAND INPUT

    will give you information about providing input data for BOSOR.

    Information is available on the following topics:

    COMMAND     DEBUG       FILES
    OVERVIEW    PROGRAMS
```

Next, please type HELP4 TOPIC, in which TOPIC stands for one of the
topics listed above.

$ HELP4 OVERVIEW
<BOSOR4>
   OVERVIEW

   Information is available on:

   ANALYSES   HOWTORUN   SAMPLERUN   SUMMARY
   EXPERT   DOCUMENTS   LOGIN

   Next, please type HELP4 OVERVIEW SUBTOPIC, in which SUBTOPIC stands
   for one of the subtopics listed above.

$ HELP4 OVERVIEW SUMMARY
<BOSOR4>
   SUMMARY of purpose of BOSOR4...

In BOSOR4 a complex, branched, stiffened shell of revolution is
treated as an assemblage of shell segments or branches, each with its
own geometry (flat, cylindrical, spherical, etc), loading, wall
construction, and material properties. In this version of BOSOR4 you
provide the input data in a "conversationally" interactive mode on a
segment-by-segment basis.

Before beginning an interactive session you should already have
decided how to divide the structure that you are analyzing into
segments and branches, and you should have available to you all
dimensions and material properties and loading. You should also
have decided what kind of analysis to perform.

You provide input data interactively. The prompting
information contains references to page numbers and figures and
tables that are contained in:

"BOSOR4--Program for stress, buckling, and vibration of complex shells
of revolution", STRUCTURAL MECHANICS SOFTWARE SERIES,Vol. I, edited by
N. Perrone and W. Pilkey, Univ. Press of Virginia,Charlottesville,
pp. 11-143 (1977). This is the user's manual for earlier BOSOR4's.

$ HELP4 OVERVIEW HOWTORUN
<BOSOR4>
   HOW TO RUN BOSOR...

      You first activate BOSOR4 commands via the command BOSOR4LOG.
      This command must be given before you do any BOSOR4 work.
      Input data are then prepared interactively via the INPUT
      command or a series of INPUT commands. Files with the name
      'NAME'.SEG;n , n = 1, 2, 3, . . .(NSEG+1), in which 'NAME' is
      a name you assign to the case and NSEG is the number of
      structural segments in the model, are then concatenated into
      a single file, 'NAME'.ALL, via ASSEMBLE. Then a batch run
      is initiated via the command BOSORALL. Plots are obtained
      via the command BOSORPLOT.

      SAMPLE RUNSTREAM:

               BOSOR4LOG   (You activate the BOSOR4 commands.)
               INPUT   (You supply data interactively. It is
                               automatically stored on 'NAME'.SEG;n)

```
                ASSEMBLE        ('NAME'.SEG;n files are concatenated into
                                 a single file, 'NAME'.ALL)
                BOSORALL        (A batch run through the BOSOR4 pre-,
                                 main-, and post-processors is initiated)
                BOSORPLOT       (Plot files *.PLV are generated)

$ HELP4 COMMANDS
<BOSOR4>
   COMMANDS

   Information is available on:

ASSEMBLE      BOSORALL      BOSORPLOT     CHECKFILE     CLEANUP
GETLIB        GETSEGS       INPUT         MODIFY
ORDERSEG      RESETUP       RESTART       BOSOR4LOG     PLOT

   Next, please type HELP4 COMMAND SUBTOPIC, in which SUBTOPIC stands
   for one of the subtopics listed above.

$ HELP4 COMMANDS MODIFY
<BOSOR4>
   MODIFY

        This command allows you to edit files with the names 'NAME'.SEG.
        You first answer questions pertaining to a particular category of
        structural information. This short interactive session produces
        a file with a name such as GEOM.QUE (for meridional geometry),
        RINGS.QUE (for discrete rings), LINELOADS.QUE (for line loads),
        CONSTRAIN.QUE (for constraint conditions), etc. Following the
        short interactive session, (which you can bypass if a suitable
        *.QUE file already exists), MODIFY automatically inserts this
        *.QUE file into the proper position in the 'NAME'.SEG file and,
        if you wish, tests the new 'NAME'.SEG file using the utility
        CHECKFILE.

MODIFY works because of certain key phrases embedded in the 'NAME'.SEG
files. These key phrases tell MODIFY where the new data should
be inserted. Because of this, MODIFY should only be used with NAME.SEG
files that have been generated from a good NAME.DOC file via the
utility GETSEGS. (GETSEGS disassembles the NAME.DOC file into NAME.SEG
files.) MODIFY may not work when you apply it to NAME.SEG files that
have been generated in some shortcut manner, such as by means copying
from other segments or otherwise using the VAX editor to generate
NAME.SEG files.

            APPENDIX 2.  BOSOR4 SAMPLE CASE EXTERNALLY PRESSURIZED SPHERICAL CAP
            ON FRICTIONLESS PLANE

Appendix 2 contains the interactive session for input data.

   $ BOSOR4LOG
Previous logical name assignment replaced

BOSOR4 COMMANDS HAVE BEEN ACTIVATED.

The BOSOR4 commands, in the general order in which
you would probably use them, are:

   HELP4        (get information on BOSOR4.)
   INPUT        (you provide segment-by-seg. input)
   ASSEMBLE     (concatenates segment data files)
```

```
BOSORALL    (batch run of pre, main, post proc.)
BOSORPLOT   (batch run for generating plot files)
RESETUP     (input for restart run, same model)
RESTART     (batch run of main and postprocessors)
CLEANUP     (delete all except for .DOC file)
GETSEGS     (generate segment files from .DOC)
MODIFY      (modify a segment file)
CHECKFILE   (check a segment file)
```

Please consult the following sources for more
information about BOSOR4:

1. HELP4 file   (type HELP4)
2. BOSOR4ST.ORY   (good idea to print this file)
3. Documents listed under HELP4 OVERVIEW DOC

$ INPUT

PROMPT FILES HAVE NOW BEEN ASSIGNED.

ENTER BOSOR4 CASE NAME: CAP

Do you want to provide data for a new structural
segment, or add data to that for an existing
structural segment (Y or N) ? : Y

AGAIN, ENTER THE BOSOR4 CASE NAME
CAP
Which segment is this?=1
Are you correcting, adding to, or checking an existing file?=N

Please provide a title (42 characters or less)...

EXT. PRESSURE ON CAP ON FRICTIONLESS PLANE
INDIC = analysis type indicator=H

 0= nonlinear axisymmetric stress (and collapse) analysis,
-2= stability determinant calculated for increasing load,
-1= bifurcation buckling with nonlinear prebuckling analysis,
 1= bifurcation buckling with linear   prebuckling analysis;
    (Actually the prebuckling analysis is the same as for
     INDIC = -1. However, the applied load is never changed
     during a case. Linear behavior is exhibited as long as
     the user applies a load that is very small compared to
     the design load.)

 2= modal vibration with axisymmetric nonlinear prestress.
 3= linear nonsymmetric stress analysis;
 4= bifurcation buckling with linear nonsymmetric prebuckling.

INDIC = analysis type indicator=-2
NPRT = output options (1=minimum, 2=medium, 3=maximum)=1
ISTRES= output control (0=resultants, 1=sigma, 2=epsilon)=0
NSEG = number of shell segments (less than 95)=1

The following input must be provided by you for each
shell segment. See p. 61 for a list of the types of
input data required. (Page numbers refer to the BOSOR4 user's
manual, "BOSOR4--Program for Stress, Buckling, and Vibration of
Complex Shells of Revolution", STRUCTURAL MECHANICS SOFTWARE SERIES-
Vol. 1, edited by N. Perrone and W. Pilkey, University Press of
Virginia, Charlottesville, VA, 1977, pp 11-143)

NMESH = number of node points (5 = min.; 98 = max.)=31
NTYPEH= control integer (1 or 2 or 3) for nodal point spacing=3

Geometry of the current segment...
NSHAPE= indicator (1,2 or 4) for geometry of meridian=2
R1   = radius at beginning of segment (see p. 66)=0
Z1   = axial coordinate at beginning of segment=0
R2   = radius at end of segment=75.16523
Z2   = axial coordinate at end of segment=34.04405
RC   = radius from axis of rev. to center of curvature=0
ZC   = axial coordinate of center of curvature=100
SROT=indicator for direction of increasing arc (-1. or +1.)=H

See Figures on p. 66
-1 means anticlockwise; +1 means clockwise.

SROT=indicator for direction of increasing arc (-1. or +1.)=-1.

Imperfection geometry....
IMP   = indicator for imperfection (0=none, 1=some)=0
NTYPEZ= control (1 or 3) for reference surface location=H

NTYPEZ = 1 means that the distance from the shell wall
leftmost surface to the reference surface varies along
the meridian. By "leftmost" we mean as we face in the
direction of increasing meridional arc length, s. See
the figure at the bottom of p. 66.

NTYPEZ = 3 means that the distance from the leftmost
surface of the wall to the reference surface is constant as we proceed along the meridian, s.

NTYPEZ= control (1 or 3) for reference surface location=3
ZVAL = distance from leftmost surf. to reference surf.=.5
Do you want to print out r(s), r'(s), etc. for this segment?=Y
NRINGS= number (max=20) of discrete rings in this segment=0
K=elastic foundation modulus (e.g. lb/in**3)in this seg.=0

The following input is related to loading of this
segment. Please see pp. 73-77 for discussion and
definitions. Also, you may wish to review pp. 58-60.
There is more discussion in the "pitfalls" section
on pp. 120-123.

There are four classes of loads:
a. mechanical line loads and/or imposed displacement
   components, applied at centroids of discrete rings;
b. thermal line loads at discrete rings;
c. pressure and tractions distributed over the surface;
d. temperature distribution through thickness and over
   surface.

In connection with mechanical line loads and/or imposed
displacement components, the word "load" is used to mean
either an imposed load or an imposed displacement.

LINTYP= indicator (0, 1, 2 or 3) for type of line loads=H

0 means none
1 means mechanical line loads and/or imposed displacements;
2 means thermal line loads only;
3 means both mechanical and thermal line loads.

Note that if LINTYP is greater than 0 there must be
discrete rings on which to "hang" the line loads and/or
imposed displacement components.
Line loads are assumed to act at the centroids of discrete
rings. They are positive as shown on page 74, bottom.
Imposed displacement components also "act" at ring centroids.
They are positive as shown on page 51, bottom (USTAR,WSTAR,CHI).

In the following input for line loads or imposed displacements...
 V(K) can mean axial load or imposed axial displacement;
      [note: positive V (load) is in opposite direction from
             positive V (imposed axial displacement USTAR)]
 S(K) can mean circ. load or imposed circ. displacement;
 HF(K) can mean radial load or imposed radial displacement;
 FM(K) can mean meridional moment or imposed rotation CHI (p.51).

LINTYP= indicator (0, 1, 2 or 3) for type of line loads=0
IDISAB= indicator (0, 1, 2 or 3) for load set A and B=H

0 means no distributed loads (no pressure or thermal loading)
1 means only distributed load set A is present
2 means only distributed load set B is present
3 means both distributed load set A and distributed load set B
  are present

Load set A is considered to be multiplied by the eigenvalue,
whereas load set B is not. Load set B is a fixed preload.

IDISAB= indicator (0, 1, 2 or 3) for load set A and B=1

Next, provide input for distributed loads in load set A.
(loads that are to be multiplied by the eigenvalue)...

SURFACE LOADS FOR LOAD SYSTEM "A"...
NLTYPE=control (0,1,2,3) for type of surface loading=H

NLTYPE= 0 means no pressure, surface traction, or
          temperature distribution on this shell segment.
NLTYPE= 1 means pressure and/or surface traction, but
          no temperature distribution on this segment.
NLTYPE= 2 means temperature distribution, but no pressure
          or surface traction.
NLTYPE= 3 means both pressure and temperature.

NLTYPE=control (0,1,2,3) for type of surface loading=1
NPSTAT= number of meridional callouts for surface loading=H

Minimum value is NPSTAT = 2, corresponding to callout points
at the beginning and at the end of the segment. Maximum
value is NPSTAT = 20

NOTE: The first at last points along the meridian must be
      included as callouts.

NPSTAT= number of meridional callouts for surface loading=2
NLOAD(1)=indicator for meridional traction (0=none, 1=some)=0
NLOAD(2)=indicator for circumferential traction=0
NLOAD(3)=indicator for normal pressure   (0=none, 1=some)=1
PN(i)   = normal pressure (p.74) at ith callout, PN(1)=H

Sign convention is shown in the fig. on p. 74. Also, see

BOSOR4: Program for Stress, Stability and Vibration   49

Table A4 and p. 85. The total pressure is PN(i)*g(theta),
where g(theta) is a Fourier series defined below. In the
figure on p. 74 the loads PT, PC, and PN are called p1, p2, p3,
respectively.

PN(i) = normal pressure (p.74) at ith callout, PN(1)=-1.
PN(i) = normal pressure (p.74) at ith callout, PN(2)=-1.

NTYPE = control for meaning of loading callout (2=z, 3=r)=H

See pp. 69 for further discussion and examples.
NTYPE = 2 means callouts for meridional variation of
        surface traction and pressure will be axial
        coordinates;
NTYPE = 3 means callouts will be radial coordinates.

NTYPE = control for meaning of loading callout (2=z, 3=r)=3
R(I) = radial coordinate of Ith loading callout, r(1)=0
R(I) = radial coordinate of Ith loading callout, r(2)=75.16523

Wall construction input follows...

NWALL=index (1, 2, 4, 5, 6, 7, 8) for wall construction=H

NWALL = 1 means general C(i,j) (see p.90)
NWALL = 2 means monocoque isotropic
NWALL = 4 means fiberwound, layered, constant thickness
NWALL = 5 means layered orthotropic, variable thickness
NWALL = 6 means corrugated (corrugations run axially)
NWALL = 7 means semi-sandwich axially corrugated, that is
        a smooth sheet is fastened to a corrugated sheet
NWALL = 8 means layered orthotropic with temperature-
        dependent material properties, variable thickness

Smeared stiffeners may be added to any of these types.
The smeared stiffeners may be either or both rings and
stringers.

NWALL=index (1, 2, 4, 5, 6, 7, 8) for wall construction=2

Input for monocoque, isotropic wall construction...

E   = Young's modulus for skin=10000000.
U   = Poisson's ratio for skin=.3
SM   =mass density of skin (e.g. alum.=.00025 lb-sec**2/in**4)=0
ALPHA = coefficient of thermal expansion=0
NRS   = control (0 or 1) for addition of smeared stiffeners=0
NSUR = control for thickness input (0 or 1 or -1)=H

NSUR = 0 means reference surface is middle surface
         (We will not need to provide thickness, since
          we have already provided ZVAL, the distance
          from the leftmost surface to the ref. surf.)

NSUR = 1 means the reference surface is the outer or
          rather the rightmost surface. Again, we do
          not need to provide input for the thickness,
          since ZVAL is the same as the thickness in
          this case.

NSUR =-1 means that the reference surface is arbitrarily

```
                  located with respect to the leftmost surface (It
                  might be the leftmost surface itself). Therefore,
                  you will have to provide additional input data
                  to establish the wall thickness.

NSUR    = control for thickness input (0 or 1 or -1)=0
Do you want to print out ref. surf. location and thickness?=N
Do you want to print out the C(i,j) at meridional stations?=N
Do you want to print out distributed loads along meridian?=N


Directory DRC1:[BUSHNELL]

CAP.SEG;1              8/8        22-OCT-1984 17:30

Total of 1 file, 8/8 blocks.
Want to add more structural segments now (Y or N) ? : N
Have you supplied data for all structural segments? : Y

Next, give global input and input for constraint
conditions....

Do you want to supply these data now (Y or N) ? : Y
Previous logical name assignment replaced

AGAIN, ENTER THE BOSOR4 CASE NAME
CAP
How many segments in the structure?=1
INDIC = analysis type indicator=-2
Are you correcting, adding to, or checking an existing file?=N
NLAST = plot options  (-1=none, 0=geometry, 1=u,v,w)=0
```

Your structure may contain segments that are very short
compared to the whole model being analyzed here. This detail
will not show up well in plots of the entire undeformed and
deformed structure. Therefore you may wish to get expanded
plots of these regions. Please identify these regions by
segment number and give a magnification factor for each region.
Note that the magnification factor must be an integer.
The center of the expanded plot will be at the first point of
the segment so identified. The extent of structure plotted
will of course depend on the magnification factor you choose.

```
Are there any regions for which you want expanded plots?=N
NOB   = starting number of circ. waves (buckling analysis)=6
NMINB = minimum number of circ. waves  (buckling analysis)=2
NMAXB = maximum number of circ. waves  (buckling analysis)=15
INCRB = increment in number of circ. waves (buckling)=1
NVEC  = number of eigenvalues for each wave number=1
```

Next, please provide factors P and DP, TEMP and DTEMP, which
are multipliers for the pressure, surface traction, and
temperature distributions in load system "A". Note that these
multipliers are applied only to load system "A". They are not
applied to load system "B". (Load system "A" represents the
"eigenvalue" load system. Load system "B" is constant
throughout the case.)

        P   = pressure or surface traction multiplier=H

The factor P is applied only to the distributed mechanical
loads in load system "A". For example, if INDIC is less than
three (axisymmetric loading), the normal pressure along Segment

BOSOR4: Program for Stress, Stability and Vibration 51

No. i for load system "A" is given in the first load step by:

  pressure = P*PN(j)   j = 1, 2, ...NMESH(i)

in which PN(j) is the meridional pressure distribution.

See pp. 58-60 for further discussion of loading parameters.

P   = pressure or surface traction multiplier=200
DP   = pressure or surface traction multiplier increment=100
TEMP   = temperature rise multiplier=0
DTEMP = temperature rise multiplier increment=0
Number of load steps=20
OMEGA = angular vel. about axis of revolution (rad/sec)=0
DOMEGA = angular velocity increment (rad/sec)=0
How many segments in the structure?=1

Four kinds of constraint conditions exist in BOSOR4:

1. constraints to ground (e.g. boundary condtions)
2. juncture compatibility conditions
3. regularity conditions at poles (where radius r = 0)
4. constraints to prevent rigid body displacements

See the fig. on p. 54, for example. There is a constraint to
ground (boundary condition) at Segment 8, Point 8; there are
several juncture conditions (e.g. Seg. 2, Pt. 1 is connected
to Seg. 1, Pt. 9); there are several poles (e.g. Seg. 1,
Pt. 1). Note that if a shell is not anywhere attached to
ground, such as is the case for the example shown on p. 57,
the user must choose a node at which to prevent rigid body
motion. This node is to be chosen in the section below where
the user is asked about constraints to ground. In a section
following the "constraints-to-ground" section, the user will
be asked to provide specific data for preventing rigid body
motion. Types of rigid body motion are shown on p. 56. An
example of appropriate input data is listed on p. 57, bottom.

  CONSTRAINT CONDITIONS FOR SEGMENT NO. ISEG =   1
Number of poles (places where r=0) in SEGMENT=1
IPOLE = nodal point number of pole, IPOLE(1)=1
At how many stations is this segment constrained to ground?=1
INODE = nodal point number of constraint to ground, INODE(1)=31
IUSTAR=axial displacement constraint (0 or 1 or 2)=1
IVSTAR=circumferential displacement(0=free,1=0,2=imposed)=0
IWSTAR=radial displacement(0=free,1=constrained,2=imposed)=0
ICHI=meridional rotation (0=free,1=constrained,2=imposed)=0
D1   = radial component of offset of ground support=0
D2   = axial component of offset of ground support=0
Is this constraint the same for both prebuckling and buckling?=Y
Is this segment joined to any lower-numbered segments?=N

It may be necessary to provide additional constraint to
ground in order to prevent rigid body motion. All possible
types of rigid body motion are shown on p. 56. Rigid body
motion corresponds to n = 0 or n = 1 circumferential waves.
There is no rigid body component for any harmonic with n
greater than or equal to 2. For modal vibration problems rigid
body motion need be prevented only if the structure is loaded.

Given existing constraints, are rigid body modes possible?=N
Do you want to list output for segment(1)=Y

Do you want to list forces in the discrete rings, if any?=N

Directory DRC1:[BUSHNELL]

CAP.SEG;2   6/8   22-OCT-1984 17:40
CAP.SEG;1   8/8   22-OCT-1984 17:30

Total of 2 files, 14/16 blocks.

If you have completed input for all structural
segments and for the constraint conditions, next
give the command ASSEMBLE

$ ASSEMBLE
Enter BOSOR case name: CAP
Enter number of segments in the structure: 1
 DELETE-W-FILNOTDEL, error deleting DRC1:[BUSHNELL]FORO*.*;*
 -RMS-E-FNF, file not found
 DELETE-W-FILNOTDEL, error deleting DRC1:[BUSHNELL]CAP.BLK;*
 -RMS-E-FNF, file not found

Don't worry about above error messages, if any.

You have the following files with the name
CAP.ALL :
  No files found.

Any old CAP.ALL files that you want to save (Y or N)?: N
 DELETE-W-FILNOTDEL, error deleting DRC1:[BUSHNELL]CAP.ALL;*
 -RMS-E-FNF, file not found

You have the following files with the name
CAP.SEG :

Directory DRC1:[BUSHNELL]

CAP.SEG;2   6/8   22-OCT-1984 17:40
CAP.SEG;1   8/8   22-OCT-1984 17:30

Total of 2 files, 14/16 blocks.

Do you have exactly 2 CAP.SEG files (Y or N)? : Y
Are the CAP.SEG files in the correct order (Y or N)?: Y
Do you want to delete the CAP.SEG files? (Y or N): N
 CAP.SEG;n files are now
            assembled into the file CAP.ALL
You now have the following files with the name
CAP.* :
Directory DRC1:[BUSHNELL]
CAP.ALL;2   13/16   22-OCT-1984 17:45
CAP.SEG;2   6/8   22-OCT-1984 17:40
CAP.SEG;1   8/8   22-OCT-1984 17:30
Total of 3 files, 27/32 blocks.
Now you can give the command BOSORALL

 $ BOSORALL
ENTER BOSOR4 CASE NAME: CAP
WHAT DISK:[DIRECTORY.SUBDIR] FOR YOUR I/O?: DRC1:[BUSHNELL.MESH]
WHAT DISK:[DIRECTORY.SUBDIR] CONTAINS BOSOR4?: DRC1:[BUSHNELL.BOSOR4]
F(AST),S(YS$BATCH),T(EST),N(ORMAL),OR H(UGE) QUEUE?: F

Give time, in the format 04-JUL-1984:16:00, after
which you want the batch run started. If you want
the batch run started as soon as possible, simply
hit RETURN.

RUN BATCH JOB AFTER (RETURN or DAY-MONTH-YEAR:HOUR:00):
Job 8320 entered on queue FAST

APPENDIX 3. ANNOTATED FILE, CAP.DOC, CREATED BY BOSOR4 PREPROCESSOR

```
EXT. PRESSURE ON CAP ON FRICTIONLESS PLANE
      -2         $ INDIC = analysis type indicator
       1         $ NPRT = output options (1=minimum, 2=medium, 3=maximum)
       0         $ ISTRES= output control (0=resultants, 1=sigma, 2=epsilon)
       1         $ NSEG = number of shell segments (less than 95)
       H         $
       H         $ SEGMENT NUMBER    1   1   1   1   1   1   1   1
       H         $ NODAL POINT DISTRIBUTION FOLLOWS...
      31         $ NMESH = number of node points (5 = min.; 98 = max.)( 1)
       3         $ NTYPEH= control integer (1 or 2 or 3) for nodal point spacing
       H         $ REFERENCE SURFACE GEOMETRY FOLLOWS...
       2         $ NSHAPE= indicator (1,2 or 4) for geometry of meridian
       0         $ R1    = radius at beginning of segment (see p. 66)
       0         $ Z1    = axial coordinate at beginning of segment
  75.16523       $ R2    = radius at end of segment
  34.04405       $ Z2    = axial coordinate at end of segment
       0         $ RC    = radius from axis of rev. to center of curvature
     100         $ ZC    = axial coordinate of center of curvature
  -1.000000      $ SROT=indicator for direction of increasing arc (-1. or +1.)
       H         $ IMPERFECTION SHAPE FOLLOWS...
       0         $ IMP   = indicator for imperfection (0=none, 1=some)
       H         $ REFERENCE SURFACE LOCATION RELATIVE TO WALL
       3         $ NTYPEZ= control (1 or 3) for reference surface location
   0.5000000     $ ZVAL  = distance from leftmost surf. to reference surf.
       Y         $ Do you want to print out r(s), r'(s), etc. for this segment?
       H         $ DISCRETE RING INPUT FOLLOWS...
       0         $ NRINGS= number (max=20) of discrete rings in this segment
       0         $ K=elastic foundation modulus (e.g. lb/in**3)in this seg.
       H         $ LINE LOAD INPUT FOLLOWS...
       0         $ LINTYP= indicator (0, 1, 2 or 3) for type of line loads
       H         $ DISTRIBUTED LOAD INPUT FOLLOWS...
       1         $ IDISAB= indicator (0, 1, 2 or 3) for load set A and B
       H         $ SURFACE LOAD INPUT FOR LOAD SET "A" FOLLOWS
       1         $ NLTYPE=control (0,1,2,3) for type of surface loading
       2         $ NPSTAT= number of meridional callouts for surface loading
       0         $ NLOAD(1)=indicator for meridional traction (0=none, 1=some)
       0         $ NLOAD(2)=indicator for circumferential traction
       1         $ NLOAD(3)=indicator for normal pressure     (0=none, 1=some)
  -1.000000      $ PN(i)   = normal pressure (p.74) at ith callout, PN( 1)
  -1.000000      $ PN(i)   = normal pressure (p.74) at ith callout, PN( 2)
       3         $ NTYPE = control for meaning of loading callout (2=z, 3=r)
       0         $ R(I)  = radial coordinate of Ith loading callout, r( 1)
  75.16523       $ R(I)  = radial coordinate of Ith loading callout, r( 2)
       H         $ SHELL WALL CONSTRUCTION FOLLOWS...
       2         $ NWALL=index (1, 2, 4, 5, 6, 7, 8) for wall construction
 0.1000000E+08   $ E     = Young's modulus for skin
  0.3000000      $ U     = Poisson's ratio for skin
       0         $ SM =mass density of skin (e.g. alum.=.00025 lb-sec**2/in**4)
       0         $ ALPHA = coefficient of thermal expansion
       0         $ NRS   = control (0 or 1) for addition of smeared stiffeners
```

```
      0      $ NSUR    = control for thickness input (0 or 1 or -1)
  N          $ Do you want to print out ref. surf. location and thickness?
  N          $ Do you want to print out the C(i,j) at meridional stations?
  N          $ Do you want to print out distributed loads along meridian?
  H          $
  H          $ GLOBAL DATA BEGINS...
      0      $ NLAST = plot options  (-1=none, 0=geometry, 1=u,v,w)
  N          $ Are there any regions for which you want expanded plots?
      6      $ NOB   = starting number of circ. waves (buckling analysis)
      2      $ NMINB = minimum number of circ. waves  (buckling analysis)
     15      $ NMAXB = maximum number of circ. waves  (buckling analysis)
      1      $ INCRB = increment in number of circ. waves (buckling)
      1      $ NVEC  = number of eigenvalues for each wave number
    200      $ P     = pressure or surface traction multiplier
    100      $ DP    = pressure or surface traction multiplier increment
      0      $ TEMP  = temperature rise multiplier
      0      $ DTEMP = temperature rise multiplier increment
     20      $ Number of load steps
      0      $ OMEGA = angular vel. about axis of revolution (rad/sec)
      0      $ DOMEGA = angular velocity increment (rad/sec)
  H          $ CONSTRAINT CONDITIONS FOLLOW....
      1      $ How many segments in the structure?
  H          $
  H          $ CONSTRAINT CONDITIONS FOR SEGMENT NO.    1   1   1   1
  H          $ POLES INPUT FOLLOWS...
      1      $ Number of poles (places where r=0) in SEGMENT( 1)
      1      $ IPOLE = nodal point number of pole, IPOLE( 1)
  H          $ INPUT FOR CONSTRAINTS TO GROUND FOLLOWS...
      1      $ At how many stations is this segment constrained to ground?
     31      $ INODE = nodal point number of constraint to ground, INODE( 1)
      1      $ IUSTAR=axial displacement constraint (0 or 1 or 2)
      0      $ IVSTAR=circumferential displacement(0=free,1=0,2=imposed)
      0      $ IWSTAR=radial displacement(0=free,1=constrained,2=imposed)
      0      $ ICHI=meridional rotation (0=free,1=constrained,2=imposed)
      0      $ D1    = radial component of offset of ground support
      0      $ D2    = axial component of offset of ground support
  Y          $ Is this constraint the same for both prebuckling and buckling?
  H          $ JUNCTION CONDITION INPUT FOLLOWS...
  N          $ Is this segment joined to any lower-numbered segments?
  H          $ RIGID BODY CONSTRAINT INPUT FOLLOWS...
  N          $ Given existing constraints, are rigid body modes possible?
  H          $ "GLOBAL3" QUESTIONS (AT END OF CASE)...
  Y          $ Do you want to list output for segment( 1)
  N          $ Do you want to list forces in the discrete rings, if any?
```

BOSOR5—PROGRAM FOR BUCKLING OF COMPLEX, BRANCHED SHELLS OF REVOLUTION INCLUDING LARGE DEFLECTIONS, PLASTICITY AND CREEP

D. Bushnell

Lockheed Applied Mechanics Laboratory, Department 93-30, Building 255, 3251 Hanover Street, Palo Alto, California 94304, USA

ABSTRACT

BOSOR5 performs axisymmetric collapse and nonsymmetric bifurcation buckling including elastic-plastic material behaviour and creep. It does not supercede BOSOR4, as it has no modal vibration capability or linear nonsymmetric stress analysis capability. It will handle segmented or branched, multimaterial, stiffened shells. The wall may be layered. Smeared stringers and smeared or discrete rings are permitted to go plastic. Only static analysis is performed by BOSOR5. The strategy for solution of the nonlinear prebuckling problem is such that the user obtains reasonably accurate answers even if very large load or time steps are used. This strategy is based on a subincremental iteration method in which the size of the subincrement is automatically determined so that the change in stress is less than a certain prescribed percentage of the effective stress. Discrete rings of arbitrary cross section are considered to be assemblages of thin rectangular elements. The input is interactive as with BOSOR4.

THEORETICAL BACKGROUND AND PROGRAM OVERVIEW

BOSOR5 is based on finite difference energy minimization; trigonometric variation is assumed for the circumferential variable; Newton's method is used to solve nonlinear prestress equilibrium; inverse power iterations with spectral shifts are used for eigenvalue extraction; Lagrange multipliers are used for constraint conditions; BOSOR5 is based on thin shell theory.

BOSOR5 has been widely used since 1974. In 1983 extensive user-friendly features were added to the VAX version to make provision of input data easy and reliable. An interactive input session generates a completely annotated file that can be used for documentation and for input for future similar cases. A MODIFY utility makes updating a case much easier than before. These features are essentially the same as those described in connection with BOSOR4. (See the paper on BOSOR4 in Volume 2 for details.)

The complex segmented and branched shells of revolution may have various meridional geometry, wall construction, boundary conditions, ring reinforcements, stringer reinforcements, and types of loading. Pressure and surface traction may vary along the meridian; temperature may vary along the meridian and through the thickness. Line loads may be applied at discrete ring centroids. Each load may have its own quasi-static variation in time, so that sequential loadings, such as a thermal cycle followed by an external pressure, may be applied. In this way fabrication effects followed by service loads can be simulated with BOSOR5. Examples of this are given below. All loads must be axisymmetric.

BOSOR5 is currently in widespread use and is maintained by the developer. Notices of bugs found are distributed to all known users. BOSOR5 has been thoroughly checked out by comparisons with other known solutions, tests, and by extensive use at many different institutions the world over for about 10 years. The characteristics and status of BOSOR5 are similar to those of BOSOR4. Therefore, the reader is referred to Table 1 in the paper on BOSOR4 for details on restrictions, language, documentation, I/O, hardware, size, availability, and maintenance of BOSOR5.

FIELD OF APPLICATION

BOSOR5 performs the following analyses:

1. a nonlinear stress analysis for axisymmetric behaviour of axisymmetric shell systems (moderately large deflections, elastic-plastic, creeping material). Axisymmetric collapse is a special case of this type of analysis.

2. an eigenvalue analysis in which the eigenvalues represent buckling loads of axisymmetric shell systems subjected to axisymmetric loads. Eigenvectors correspond to axisymmetric and to nonaxisymmetric buckling modes.

BOSOR5 will handle segmented and branched shells with the same geometries as those handled by BOSOR4. (See Fig. 1 in the paper on BOSOR4.)

ASSUMPTIONS, METHOD, DISCRETIZATION, USER-FRIENDLY FEATURES OF BOSOR5

The governing assumptions on which BOSOR5 is based are the same as those on which BOSOR4 is based, except that the material can creep and can go plastic (with elastic unloading). The plasticity model is von Mises yield and associated flow law with isotropic strain hardening. Deformation theory is used for the in-plane shear modulus, which is needed in the analysis governing nonaxisymmetric bifurcation buckling. The strains are assumed to be small.

The method and discretization scheme are the same as in the case of BOSOR4, except that the presence of nonlinear and irreversible material behaviour necessitates the use of the principle of virtual work rather than the principle of minimum potential energy; there is no kinetic energy involved, since BOSOR5 handles only statics problems; and a double-iteration loop is required during the prebuckling analysis phase because both nonlinear geometric (moderately large deflection) and nonlinear material behaviours are present.

Details of the analysis are presented in references 1-4.

The discretization scheme in BOSOR5 is identical to that in BOSOR4. Figure 2 in the paper on BOSOR4 gives details.

The user-friendly features described in the paper on BOSOR4 also apply to BOSOR5. Please see Tables 3, 4, and 5 in the paper on BOSOR4 for details.

BOSOR5: Program for Buckling of Complex, Branched Shells of Revolution

EXAMPLES OF APPLICATION

DETAILED EXAMPLE CASE: The style of input and files generated with use of BOSOR5 are so similar to those generated with use of BOSOR4 that no such example will be repeated here. Please see Tables 4 and 5 of the paper on BOSOR4.

However, BOSOR5 input/output are a bit different from BOSOR4 I/O. With BOSOR5 the user executes the pre- main-, and postprocessors explicitly. (With BOSOR4 these three processors are all executed by the user's typing the one command BOSORALL). The explicit execution of each processor in BOSOR5 is better because more computer time is required to solve problems in which both geometric and material nonlinearity are present, and loading history is often important because of path dependence in problems involving plasticity. Therefore, the user generally wants to be able to interact with the analysis more often than is the case with BOSOR4, in which only geometric nonlinearity is present.

With BOSOR5, data that determine the state of the plastically deformed shell at each load step are saved, so that the user may restart the mainprocessor analysis at any load step for which this state has been determined in a previous run. The nature of nonlinear problems treated with BOSOR5 makes frequent use of the restart feature the rule rather than the exception that the use of this feature tends to be with BOSOR4.

BUCKLING OF INTERNALLY PRESSURIZED SHELLS: Ellipsoidal and torispherical heads with internal pressure can buckle because the material in the knuckle region is drawn in toward the axis of revolution as the internal pressure is increased. This deformation is displayed in Fig. 1. The material in the knuckle region is therefore under a biaxial stress field that is tensile in the meridional direction and compressive in the circumferential direction. The buckling mode consists of wrinkles in the knuckle region, as shown in Fig. 2. BOSOR5 calculates the axisymmetric prebuckled state and the lowest bifurcation pressure. Both geometric and material nonlinearity must be included to solve problems of this type accurately. Extensive comparisons with tests performed at the University of Liverpool, the University of Manchester, and in France are given in references 8, 9 and 10. In reference 10 this interesting problem is described in detail.

BUCKLING OF A WATER TANK: In 1972 in Belgium a large water tank collapsed upon being filled for the first time. The geometry of the tank is shown in Fig. 3(a). Failure appeared to be due to meridional buckles that formed in the conical region represented by Segment 9 in Fig. 3(b). This region was subjected to high meridional compression combined with circumferential tension, just the opposite of the biaxial stress state in the internally pressurized ellipsoidal head shown in Fig. 2. Accordingly, the buckles were long in the circumferential direction and short in the meridional direction, in contrast to the wrinkles displayed in Fig. 2. Figure 3(c) shows the axisymmetric prebuckling deformation of the tank predicted by BOSOR5 at the predicted bifurcation buckling load factor, $\lambda = 1.8$ times the load present at the moment of collapse. The prediction is higher than the actual because no allowance is made in the BOSOR5 model for geometric imperfections or welding prestresses. Figure 3(d) shows a redesigned tank and predicted axisymmetric deformations at load factors $\lambda = 1.0$ and at the predicted bifurcation buckling load factor, $\lambda = 2.65$.

AXISYMMETRIC COLLAPSE OF ROCKET BOOSTER STAGE: Figure 4 shows the rather complex rocket interstage geometry. The basically cylindrical shell is under uniform axial compression. Figure 5 displays the multi-segment model treated with BOSOR5. Plastic collapse with increaseing axial compression V occurs because of the large amount of meridional bending caused by the inward excursion of the axial load path in the region of the joint at Station 176. This rocket interstage was tested and failed at a load within 1% of the critical load predicted with BOSOR5.

MODELING FABRICATION EFFECTS WITH BOSOR5: The BOSOR5 program has been used for calculation of bifurcation buckling of cold-bent and welded ring-stiffened

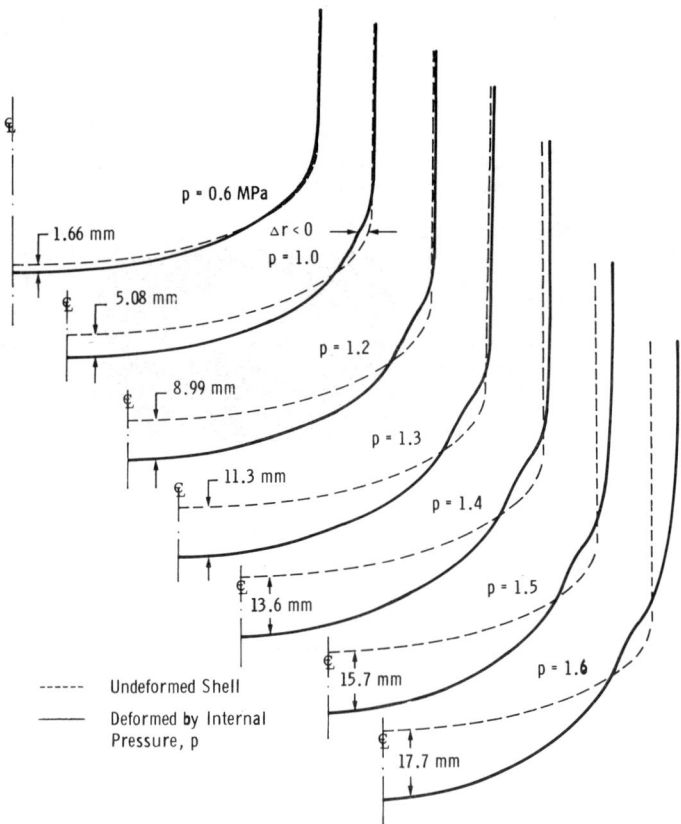

Fig. 1. Axisymmetric deformations predicted by BOSOR5 for an internally pressurized ellisoidal pressure vessel head. Elastic-plastic bifurcation buckling is caused by the relatively narrow band of circumferential compression that occurs in the knuckle region where Δr is less than zero. Accurate predictions of bifurcation buckling require inclusion of both nonlinear geometric and material behaviour in this case

cylindrical shells under external pressure. Residual stresses and deformations from cold bending a flat sheet into a cylindrical shell and subsequently welding rings to it can be incorporated into the model of buckling under service loads by introduction of these fabrication processes as functions of a time-like parameter, "time", which insures that the material in the analytical model experiences the proper sequence of loading prior to and during application of the service loads. The cold-bending process is first simulated by a thermal loading cycle in which the temperature varies linearly through the shell wall thickness, initially increasing in "time" to simulate cold bending around a die of radius R_o and then decreasing in "time" to simulate springback to a final somewhat larger design radius R. The welding process is subsequently simulated by the assumption that the material in the immediate neighborhoods of the welds is cooled below the ambient temperature by an amount that leads to weld shrinkage amplitudes typical of those observed in tests. After these two fabrication processes have been simulated, the service load (e.g. external pressure) is applied in increasing

BOSOR5: Program for Buckling of Complex, Branched Shells of Revolution

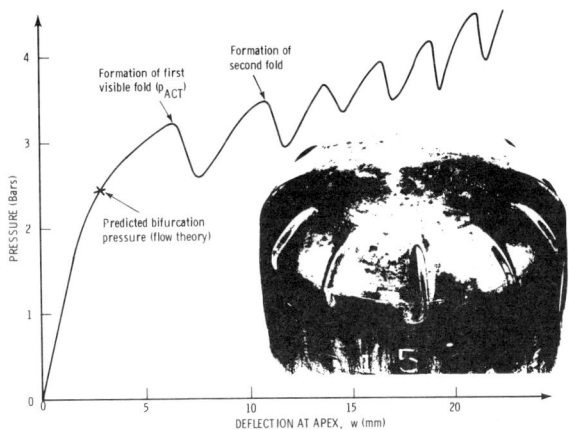

Fig. 2. Pressure-deflection curve for internally pressurized ellipsoidal pressure vessel head. BOSOR5 predicts the behaviour up to the bifurcation point. Photograph shows a head in the far postbuckled state. BOSOR5 does not predict postbifurcation behaviour

steps until buckling is detected.

Figure 6 shows two nominally identical ring-stiffened cylindrical shells tested under external pressure. The specimen on the left was fabricated by cold bending the shell and then welding machined ring stiffeners to it. That on the right was carefully machined. In the BOSOR5 analysis the machined specimen is predicted to fail at 3.723 Mpa (540 psi), precisely in agreement with the test. Simulation of only the cold-bending process leads to a prediction of p_{cr} = 3172 MPa (460 psi). Simulation of both cold bending and welding does not change this result. The predicted radial shrinkage due to welding is maximum at the ring stiffeners and minimum midway between rings, a mode that converts the straight-walled cylinder into a sort of "caterpillar". The welding process apparently has little influence on the predicted buckling pressure in this case becase of two counteracting effects: the residual welding stresses weaken the shell but the "caterpillar" type residual deformations strengthen it.

The fabricated specimen failed at a pressure about 15% below the prediction of p_{cr} = 460 psi. The discrepancy could be caused by initial geometric imperfections, the presence of the Bauschinger effect, which was not included in the BOSOR5 model, and residual stresses and other nonuniformities present in the sheet from which the cylinder was fabricated.

Figure 7 demonstrates the simulation with BOSOR5 of fabrication processes that precede application of the service load. Figure 8 shows an example from reference 11 in which the residual stresses predicted by BOSOR5 agree with those measured and calculated by Queener and DeAngelis for a cold-formed aluminium cylindrical shell. Figure 9 shows predicted residual deformation patterns in cylindrical

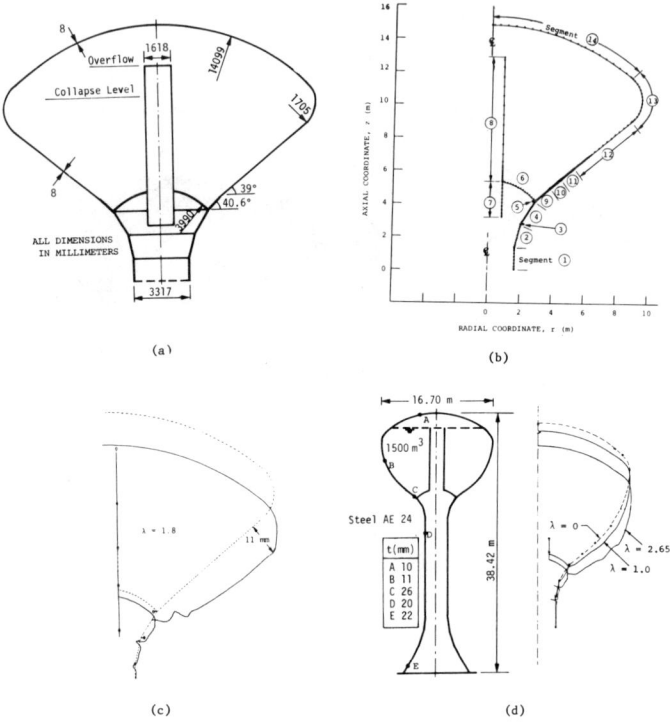

Fig. 3. Large steel water tanks and BOSOR5 models of them: (a) Geometry of original tank that failed in 1972; (b) BOSOR5 segmented and branched model of this tank; (c) BOSOR5 prediction of axisymmetric deformation just before predicted collapse; (d) Redesigned tank and deformations predicted by BOSOR5

shells caused by welding rings to the inner surface or to the outer surface of the cylindrical shell.

WORLDWIDE USE OF BOSOR5

Various versions of BOSOR5 have been in use worldwide since 1974. Table 6 in the paper on BOSOR4 lists many organizations that over the years have used BOSOR5 (and the elastic buckling program BOSOR4). Types of problems solved with the BOSOR programs are indicated.

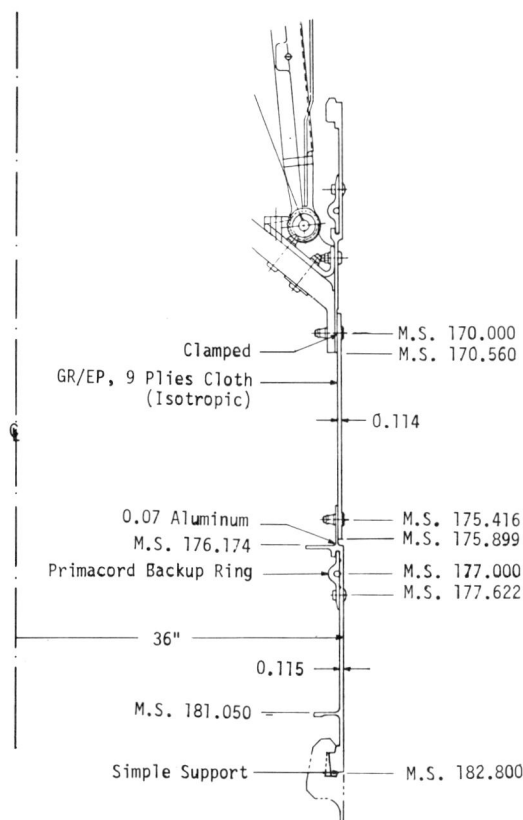

Fig 4. Complex rocket interstage subjected to axial compression during launch. (Stability problem arises from the axial load path eccentricity at Station MS 176.)

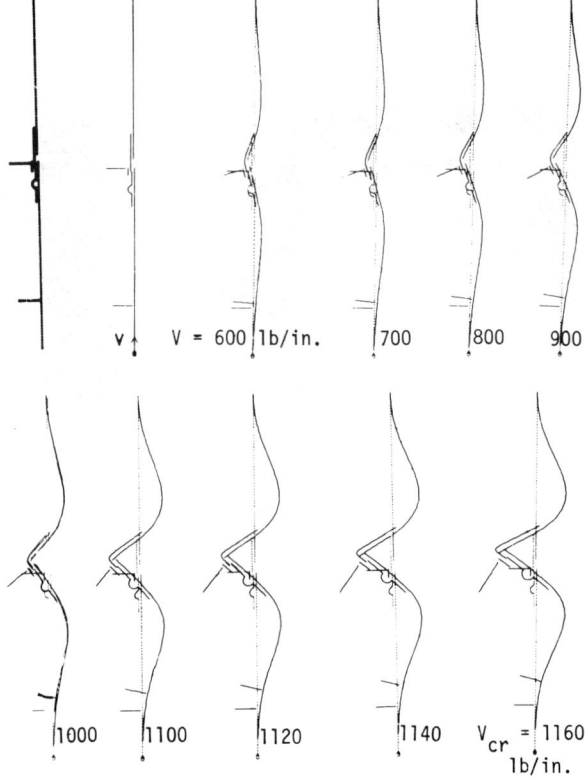

Fig. 5. Discretized, segmented BOSOR5 model and deformed profiles of the axially compressed rocket interstage with increasing axial compression V

BOSOR5: Program for Buckling of Complex, Branched Shells of Revolution   63

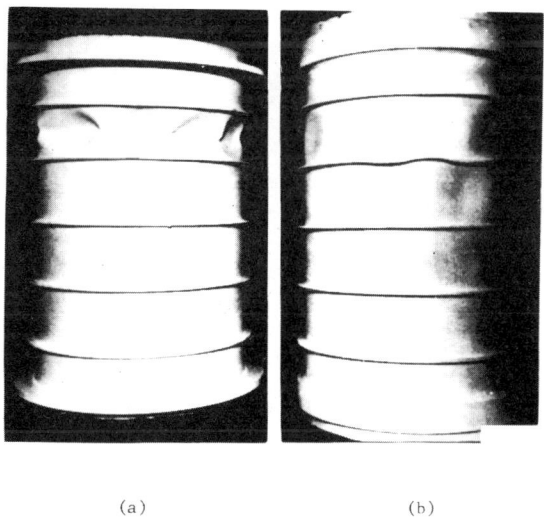

(a)   (b)

Fig. 6. Observed buckling patterns in externally pressurized,
ring-stiffened cylindrical shells: (a) The cold bent
and welded specimen on the left buckled at an external
pressure p_{cr} = 390 psi. (b) The machined specimen
buckled at p_{cr} = 540 psi

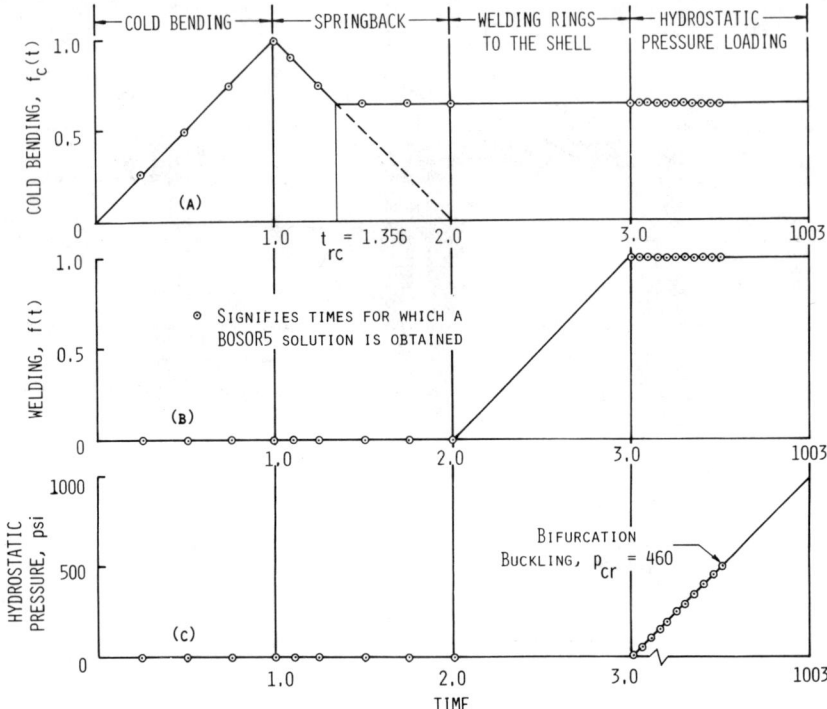

Fig. 7. Loading functions of "time" for the BOSOR5 analysis of the cold bent and welded ring-stiffened cylinder: (A) cold bending process, including springback; (B) welding the rings to the cylindrical shell; (C) application of the external hydrostatic pressure

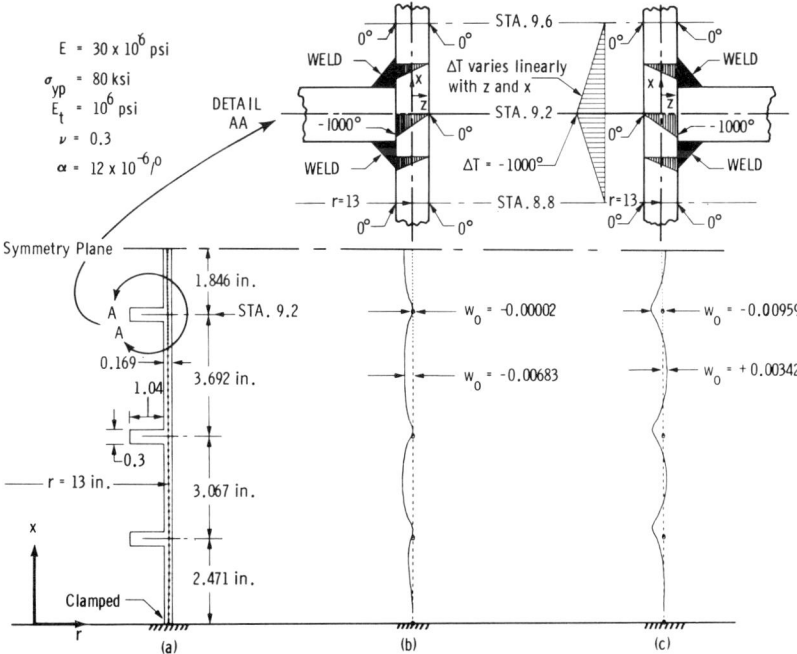

Fig. 8. Comparison of BOSOR5 results with test and theory of Queener and deAngelis for residual stress in cold bent 6061 aluminium sheet

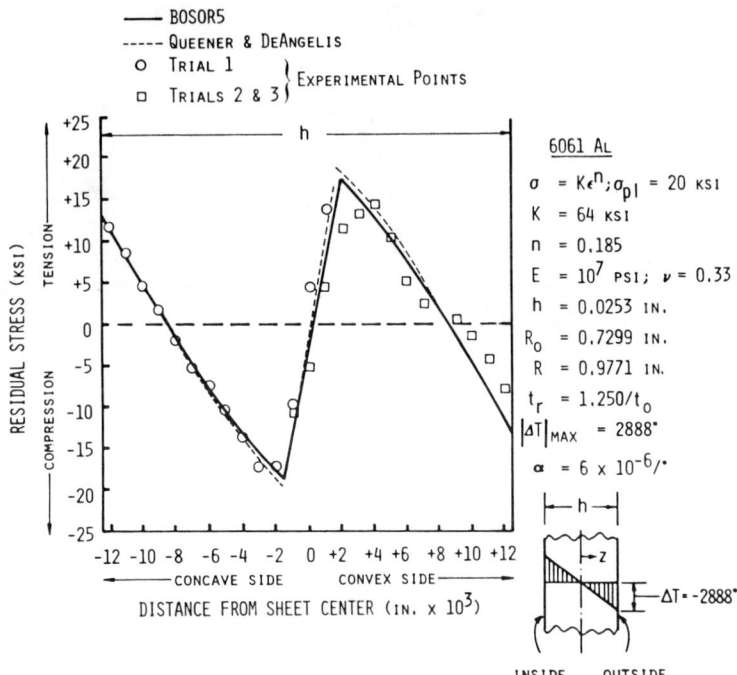

Fig. 9. Cylindrical shell with welded internal and external rings:
(a) Dimensions and BOSOR5-discretized reference surface;
(b) predicted residual deformations after welding and before
loading of a specimen with internal rings; (c) predicted
residual deformation after welding and before loading of
a specimen with external rings

REFERENCES BY D. BUSHNELL PERTAINING TO BOSOR5

1. Bushnell, D. A Computer Program for Buckling of Elastic-Plastic Complex Shells of Revolution Including Large Deflections and Creep, Vol. I: User's Manual, Input Data, LMSC-D407166; Vol. II: Test Cases, LMSC-D407167; Vol. III: Theory and Comparisons with Tests, LMSC-D407168; Lockheed Missiles and Space Co., Sunnyvale, CA., Dec., 1974.
2. BOSOR5 - Program for Buckling of Elastic-Plastic Complex Shells of Revolution Including Large Deflections and Creep. *Computers and Structures*, Vol. *6*, pp. 221-239, 1976.
3. A Strategy for the Solution of Problems Involving Large Deflections, Plasticity and Creep, *International Journal for Numerical Methods in Engineering*, Vol. *11*, pp. 683-708, 1977.
4. Bifurcation Buckling of Shells of Revolution Including Large Deflections, Plasticity and Creep, *International Journal of Solids and Structures*, Vol. *10*, pp. 1287-1305, 1974.
5. Comparisons of Test and Theory for Nonsymmetric Elastic-Plastic Buckling of Shells of Revolution (with G. D. Galletly, University of Liverpool, England), *International Journal of Solids and Structures*, Vol. *10*, pp. 1271-1286, 1974.
6. Buckling of Elastic-Plastic Shells of Revolution with Discrete Elastic-Plastic Ring Stiffeners, *International Journal of Solids and Structures*, Vol. *12*, pp. 51-66, 1976.

7. Nonsymmetric Buckling of Internally Pressurized Ellipsoidal and Torispherical Elastic-Plastic Pressure Vessel Heads, *Journal of Pressure Vessel Technology*, Vol. *99*, pp. 54-63, 1977.
8. Stress and Buckling of Internally Pressurized, Elastic-Plastic Torispherical Vessel Heads - Comparisons of Test and Theory, (with G. D. Galletly), *Journal of Pressure Vessel Technology*, Vol. *99*, pp. 39-53, 1977
9. Elastic-Plastic Buckling of Internally Pressurized Torispherical Vessel Heads, (with G. Lagae, Univ. of Ghent), *Nuclear Engineering and Design*, Vol. *48*, pp. 405-414, 1978.
10. Elastic-Plastic Buckling of Internally Pressurized Ellipsoidal Pressure Vessel Heads, *Welding Research Council Bulletin No. 267*, May 1981.
11. Effect of Cold Bending and Welding on Buckling of Ring-Stiffened Cylinders, *Computers and Structures*, Vol. *12*, pp. 291-307, 1980.
12. Plastic Buckling, from *Pressure Vessels and Piping: Design Technology, 1982 - A Decade of Progress*, edited by S. Y. Zamrik and D. Dietrich, ASME, N.Y., Book No. G00213, pp. 47-117, 1982.
13. Plastic Buckling of Various Shells, *ASME Journal of Pressure Vessel Technology*, Vol. *104*, pp. 51-72, 1984.
14. Buckling of Shells - Pitfall for Designers, *AIAA Journal*, Vol. *19*, No. 9, pp. 1183-1226, 1981.
15. Computerized Analysis of Shells - Governing Equations, *Computers and Structures*, Vol. *18*, No. 3, pp. 471-536, 1984.

ESA: ENGINEERING STRUCTURAL ANALYSIS ON PERSONAL COMPUTERS

J-P. Rammant

SCIA s.v., Belgium

ABSTRACT

This paper describes the SCIA software package for finite element analysis on microcomputers. The package has a longstanding history (starting in 1976) and has been implemented through various modules at about 350 customer sites. Its major functions are the static (linear and nonlinear) and dynamic analysis of general 2 and 3 dimensional structures. The element library covers: beams, springs, 2D solids, axisymmetric solids, plates and shells, 3D solids.

Its use is characterized through its interactive menu driven approach, with special preprocessors to facilitate the input. It was specially designed for small computers (in the beginning desktop computers) so that special algorithms were developed to overcome the restrictions of the hardware in core size and processor speed. The software has important links to CAD-CAM software, especially for steel and concrete structural design.

THEORETICAL BACKGROUND

The implementation of the finite element method on a small computer system poses some severe problems: the limitation in core size and the processor speed are the major ones. Through special algorithms (see references 1 and 2) these restrictions can be overcome.

The equation solver which we implemented follows a blocked Gaussian technique. Only three matrices of dimensions q x q (with e.g. q = 12) are in core at the same time. Its major characteristic is the necessity to have a strict bookkeeping system to read and write on disk the appropriate matrix in a minimum of time.

The speed is enhanced through the use of matrix operations, programmed in machine code. Especially the inversion, transpose and multiplication functions are utilized on several occasions. Also the speed is increased by the use of intelligent sorting techniques to find in an efficient way the corresponding matrices (e.g. element stiffness matrices) to be assembled in the global structural stiffness matrix. Searching, sorting and moving algorithms are also machine coded.

For dynamic analysis the subspace iteration method has been implemented. The practical use was extended with Stuerm sequence shifting so that a large number

of eigenvalues may be computed. The modal superposition method and the spectral
response technique follow the analysis.

The nonlinear analysis scheme follows the secans Newton Raphson method due to its
stable matrix formulation. For concrete structures the software allows the user
to examine cracking and crushing of the concrete and yielding of the steel
reinforcing bars. For frame analysis the equilibrium at the nodes is controlled
with the help of plastic moment theory: for each level of normal-force the
allowable maximum bending moment is checked and plastic hinges may form.

Since it was decided to allow for analysis of quite large structures within the
small core (64 Kbytes memory) a bandwidth reduction scheme was developed.

The major software technique to save core and to allow for large structures is
the use of a virtual data management technique. For extensive details we refer
to reference 4. In essence, the structural geometry, loading and analysis results
(displacements, nodal forces, stresses, reactions, ...) are all kept on disk in a
library system, composed of fixed pages. In the central core of the computer a
buffer array is present, where the information that is currently being treated is
kept. For the software developer and programmer that wants to interface his own
applications with the finite element software, a clear and straightforward
interface is available.

FIELD OF APPLICATION

The ESA package is divided into separate modules that threat the static, linear
and nonlinear and dynamic analysis of structures. Also heat transfer analysis
is possible.

The geometry description is provided for skeleton structures (2D and 3D trusses
and beams), 2D solids including axisymmetric models, 2D plates (with reinforcing
ribs), 3D shells and 3D solids. A separate module was developed for cold formed
steel structures (e.g. C or Z sections).

The material properties cover elasticity and plasticity (full plastic behaviour)
with possible anisotropic characteristics.

The analysis capabilities cover static, dynamic and thermal computations.
Dependent on the type of element and analysis choice several loading types are
available. In all cases automatic dead load (or mass) is taken into account.
Gravity loading, centrifugal loading, thermal loading, distributed forces,
concentrated forces are typically available.

PROGRAM DESCRIPTION

The basic theory is the finite element method, with the displacement theory as
basis. The element types are limited to a restricted choice of qualified
elements, permitting the user to not have to interfere with the difficulty of
choosing elements.

The beam element (2D and 3D) is the classical element with a third degree
polynomial displacement field. Extensions to the derivation are:

- shear force deformation correction
- eccentrically connected beam-ends
- varying cross section along the beam length
- elastically connected beam ends (partially clamped or hinged)

The 2D and 3D solid element is an improved quadrangular isoparametric element. The
initial displacement is linear in x and y. With the addition of bubble functions,
the displacement field is made quadratic. In the derivation of the stiffness matrix

numerical integration is used. It is known that one can improve the element (especially for bending behaviour) by a reduced integration technique; with controlled integration (especially for the Jacobian determinant) a full compatible element that passes the patch test is obtained. For more details see references 5 and 6.

The element has proven to be very effective for practical computations.

The 2D plate element is a triangular discrete Kirchhof element. The displacement field varies with cubic polynomes over the triangular area. Its sound behaviour, although the simple triangular shape, was proven in reference 7.

For shell analysis the above-mentioned quadrangular plane stress element and the discrete Kirchhof plate element are combined. Each element is composed of 4 triangles so that a symmetrical quadrangular element (with respect to the displacement field) is obtained.

Since the software was developed for PC (personal computer) users, interactivity was the primary objective. Without manuals, a normal first time user (with minor knowledge of the finite element method) will analyze his initial model within one hour.

Interactivity is realized with the help of menus. All input is menu driven with appropriate texts (in four languages i.e. English, German, French and Dutch) and input formats.

The presence of full mesh generators eases the use: dependent on the type of model a preprocessor is provided. For 3D trusses (grid analysis) a fast input technique is available allowing roofs to be analyzed with openings and irregular geometry without having the user inputting all 3D coordinates. For plates and shells, the preprocessor covers all stages of geometry definition: description of the form with points, lines and parts. The lines are indicated to be subdivided in a certain number of subdivisions. Through this indication the finite element mesh will be generated.

The material characteristics and the boundary conditions are defined at the stage of mesh generation input. Also the loadings are defined at the preprocessor stage. The user does not have to interfere with the generated elements and nodes.

Each program module has a graphic module to clarify the input or result interpretation. The options cover:

- plotting of geometry (2D, 3D with perspectives and projections) with representation of boundary conditions and dimensions
- plotting of loadings
- plotting of displaced structure (superposed on the undeformed form)
- numerical representation of the results (stresses)
- isoline charts for displacements and internal forces
- representation of principal stresses.

The preprocessor input is automatically checked by the program on coherence properties, e.g. for each mesh part, it is controlled if the mesh contour lines have coinciding corner points and if the subdivision numbers will yield a regular mesh. The preprocessor input is also plotted.

All graphic functions are interactive allowing the user to choose his window and sheet dimensions. Options are in the menu for plotting texts, dimensions and numbers. The isoline values are proposed by the software but can be changed manually.

A major program attraction is the link to CAD-CAM software. For general 2D and 3D (shell) analysis the input can come from a draughting package. On our microcomputer the AUTOCAD drawing package (from Autodesk Inc.) is available. The input of preprocessor points (with coordinates) can then be taken automatically from AUTOCAD. In the 3D situation the third coordinate comes from the layer number of the layer drawings (since AUTOCAD is essentially two-dimensional as yet).

The second link is provided especially towards parametric CAD packages. We have developed a steel CAD-CAM package (Steelstrac, see reference 8) that has the following modules:

- static and dynamic structural analysis
- code checking for American, Belgian, Dutch, French and German steel codes
- design of steel connections
- overview drawings and material lists
- NC modules (especially for drilling holes)

HARDWARE CONSIDERATIONS

The software is developed on an IBM PC or WANG PC or WANG 2200 computer. The PC has a MS-DOS or XENIX operation system and is based on 8086, 8080 or 80286 Intel processors (16-bit machine). A minimum configuration has 256 Kb of internal memory and a 10 Mb winchester hard disk. With this configuration a practical limit of about 800 nodes (in 2D) with a reasonable bandwidth of around 80 will be imposed by the disk

The disk space can be enhanced to 30 Mb or even larger for more complex 3D models. In most models an extra floppy diskette is present allowing an easy back-up and allowing to store the input and output data of structures.

In the near future a 32-bit processor implementation will be made available together with the UNIX operating system.

The peripherals are not restricted in types: practical all types with interfaces (Centronics or RS 232) can be connected to the computer. Special plotter drivers have been developed for TEKTRONIX, HP, BENSON, CALCOMP, HOUSTON etc.

Presently a G.K.S. (Graphical Kernel System) implementation is being developed for the graphical functions so that a standardized format will be used for all plotting.

EXAMPLES OF APPLICATION

At first we show a few tests on the isoparametric plane stress (plane strain or axisymmetric) element and on the Kirchhof plate element to show their performance. In particular a patch-test was submitted to both elements to show that the requirement is fulfilled to have monotonical convergence of the finite element results.

We consider a patch of five irregular quadrilaterals (Fig. 1) within a rectangular boundary and apply a system of forces and constraints so that a state of constant stress should be obtained (see reference 6).

Some characteristics:

E modulus = 2.07 E8 kN/m2   Poisson cf. = 0.25
thickness = 2.54 mm

Loadings: L1 = -45.72 kN   L2 = -45.72 kN
          L3 = -15.24 kN   L4 = 45.72 kN
          L5 = 45.72 kN

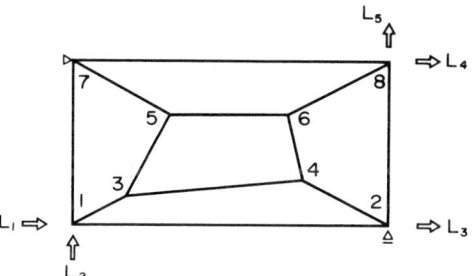

Fig. 1.

Coordinates:

Node	X	Y
1	0.0	-0.06
2	0.24	-0.06
3	0.04	-0.04
4	0.18	-0.03
5	0.08	0.02
6	0.16	0.02
7	0.0	0.06
8	0.24	0.06

Results: in each element (calculated at the centre and at the nodes) the stress state results in principal stresses

Sigma 1 = 2.0E5, Sigma 2 = 0 and Tau max = 1E5 kN/m2

Without the addition of the bubble functions and the reduced integration technique, the patch test will fail.

Another difficult test is the cantilever analysis (see reference 8). We consider 2 meshes: the first mesh has 1 element while the second mesh has 8 elements (Fig. 2).

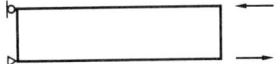 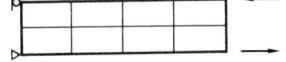

Fig. 2.

For both meshes we obtain the same results for the displacement components at the free end:

x-displ = + - 1.14116E-4 (+ at bottom, - at top)

y-displ = 1.14116 E-3

Similar tests were performed for the plate elements; as an example we show the behaviour of the element when the element aspect ratio (length of largest to smallest side) is varied. Most elements yield bad results when the shape is deformed (aspect ratios over 5). Figure 3 below shows that the discrete Kirchhof element behaves very well.

Let us briefly describe a few examples taken from industrial applications: we

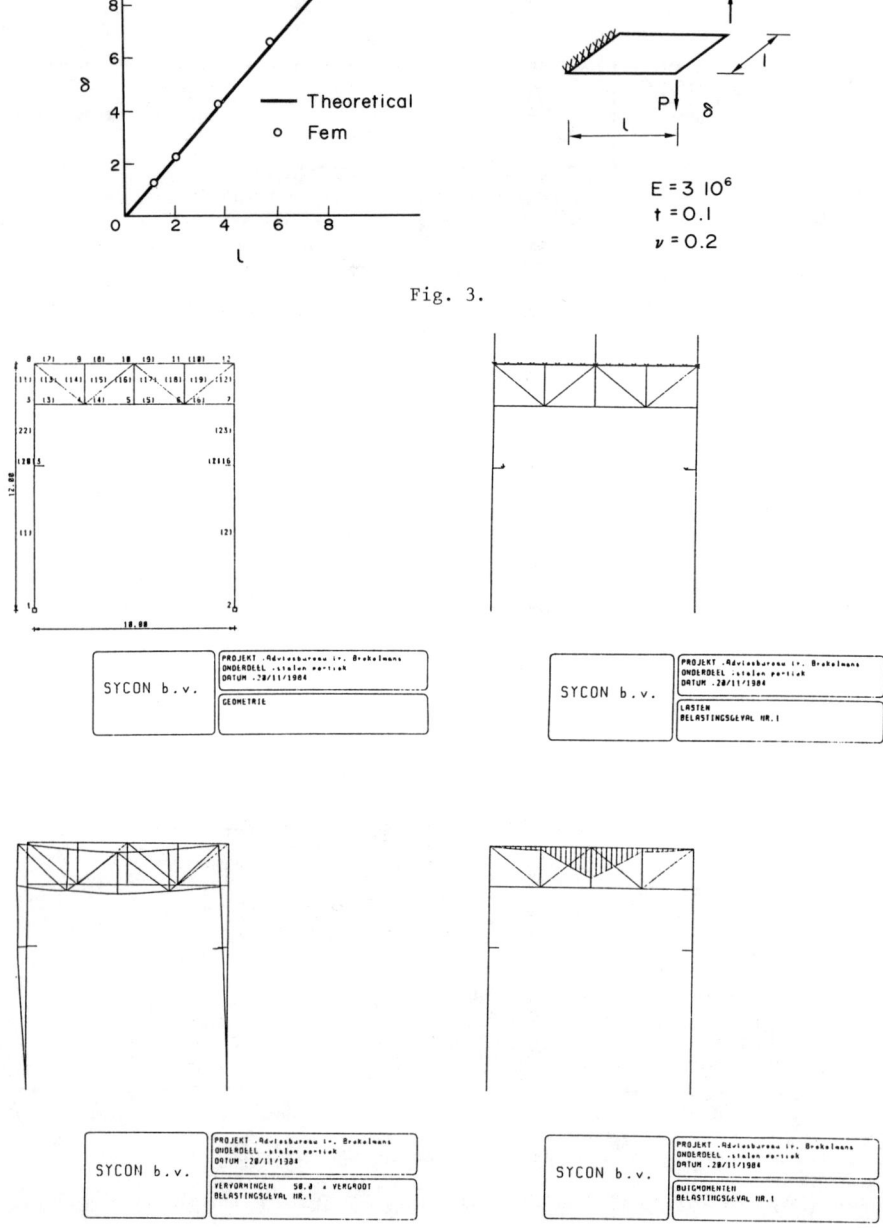

Fig. 3.

Fig. 4.

consider 4 examples, a simple portal frame structure (to show the link to CAD software), a plane stress model for a wall element, a plate structure (industrial floor) and a tank structure (shell model).

The first example is a steel structure; this type of structure is well known in industrial applications. The discretization in beam elements is very simple as one can see in Fig. 4. A picture of the loading is shown together with the deformed structure (with scale factor of 50 for the displacement values) and with the bending moments.

The standard steel sections are checked against the code and the connections are designed (bolted or welded connections). A picture of some connections and the drawn portal frame is given in Fig. 5.

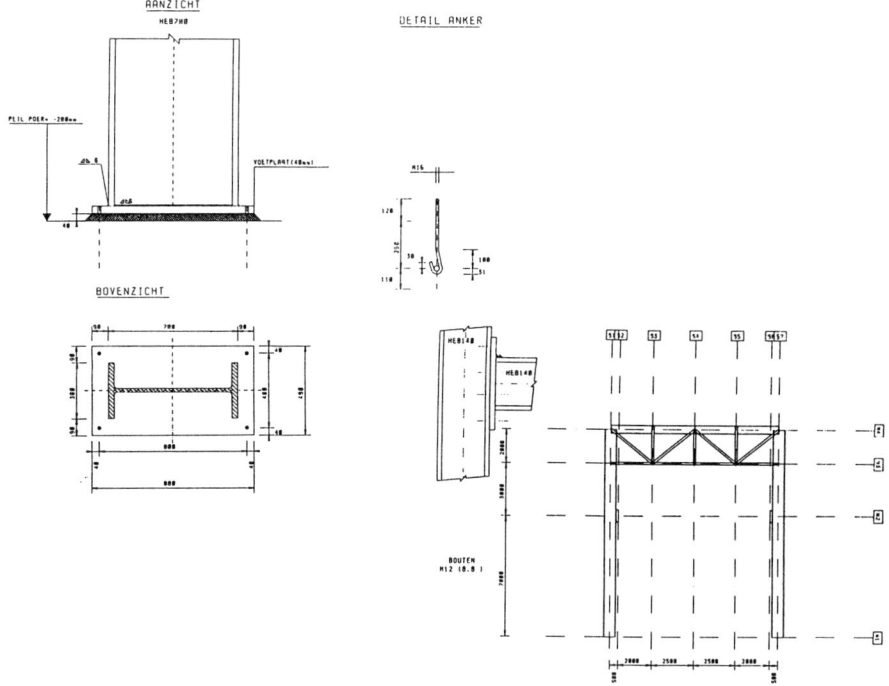

Fig. 5.

The second example is a wall element from the prefab industry. One sees two openings (one for a door and one for a window). The mesh input, generated mesh, the loading, the deformed structure and stress isolines are given in Fig. 6.

The third example is a larger model: it is a plate structure with variable thicknesses (thickness is larger at column supports). The loading, the deformed structure and isolines of the bending moments are shown in Fig. 7. The model had over 800 nodes and an automatically reduced bandwidth of about 40.

The last example is a shell structure. It is the quarter part of a tank structure resting on diaphragms. We only show the geometry that was obtained after generation by the preprocessor (Fig. 8).

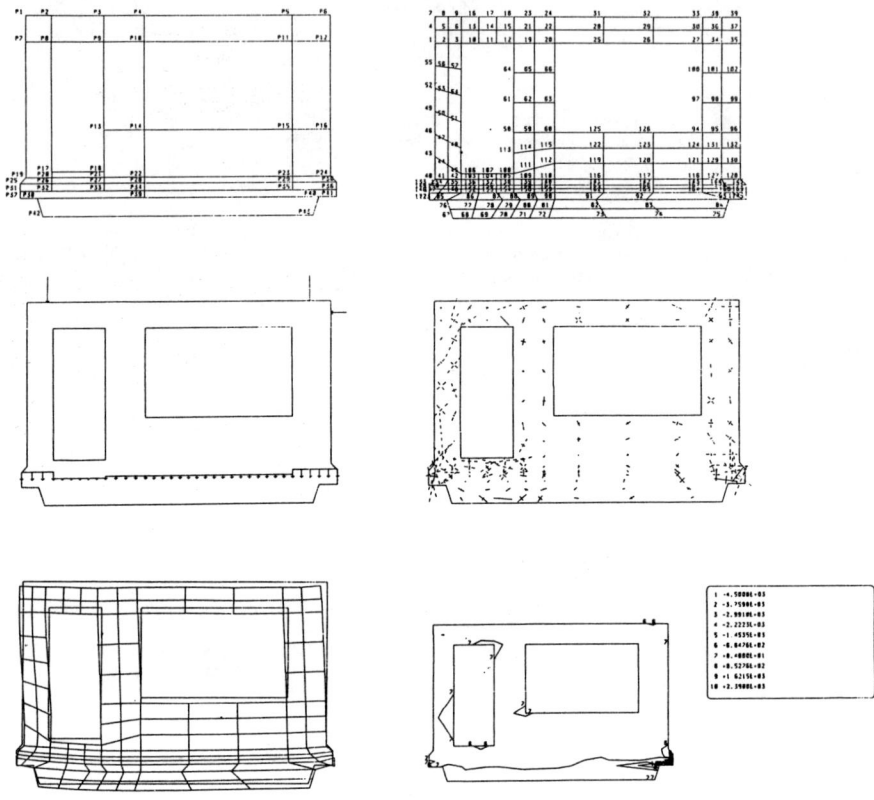

Fig. 6.

The software package that has been described in this paper is built from modules. The major finite element modules have been marketed in Europe since 1977 and over 300 users are listed as references. The majority of users are engineering bureaus together with steel construction firms and contracting firms. The size of most of the customers is small (less than 50 collaborators) although larger companies with engineering offices are also using the software. The references are mainly in Benelux, Germany, France and UK.

ESA: Engineering Structural Analysis on Personal Computers

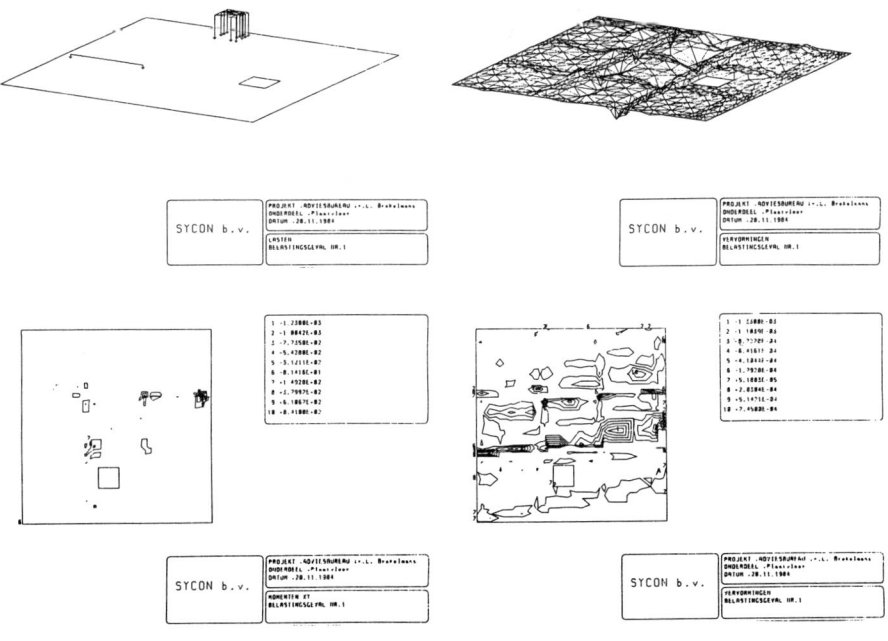

Fig. 7.

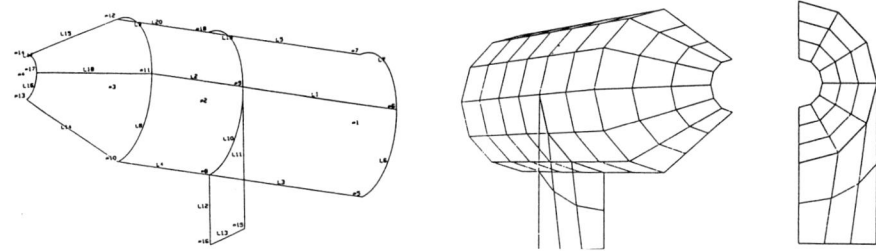

Fig. 8.

REFERENCES

1. Backx, E., J. P. Rammant and G. Schymkowitz (1978). SAP runs on a 16K desk computer. *Proc. 3rd SAP User's Conf.*, Los Angeles.

2. Rammant, J. P., E. Backx and G. Schymkowitz (1978). Case study on desk computer finite element analysis in BASIC. *2nd World Congress on F.E. Method*, Bournemouth.

3. Backx, E. and J. P. Rammant (1981). Structural dynamic interactive analysis in basic on micros, *Proc. 2nd Int. Conf Eng. Software*.

4. Backx, E. and J. P. Rammant (1983). A virtual storage data management system for F.E. analysis on micros. *ENGSOFT Conf.* London.

5. Lyons, P. (1981). An evaluation of the lower order membrane as contained in the LUCAS FEM system. *Finite Element News*, No. 4.

6. Robinson, J. and S. Blackham (1978). An evaluation of lower order membranes as contained in the ANSYS and SAP4 FEM system. *Finite Element News*, No. 2.

7. Batoz, J-L, K-J Bathe and L-W Ho (1978). A search for the optimum 3-node triangular plate bending element. *MIT Rep.* 82448-8.

8. Zienkiewicz, O. C. (1971). *The Finite Element Method in Engineering Science*. McGraw-Hill, London.

HYBRID: DETERMINATION OF NOTCH AND CRACK BOUNDARY STRESSES USING HYBRID ELEMENTS WITH THE PROGRAM SYSTEM HYBRID

Drumini, Rudi and Carmine

T. H. Karlsruhe, D75 Karlsruhe, Federal Republic of Germany

ABSTRACT

The program system renders an effective FEM-analysis of plane stress concentration problems possible at notches and cracks. The boundary stresses along the length of cracks and along curved notches can be computed by applying the hybrid method. In regions of stress concentrations, hybrid elements are used, the stress interpolation functions of which satisfy exactly the boundary conditions and the equilibrium condition. In regions of low stress gradients, the mesh is prepared by triangular elements, which are compatible with the hybrid elements used at notch and crack surfaces.

A mesh-generator executes the automatic renumbering of the nodes and elements. Boundary stress and displacement conditions can be put in by input data. The curved boundaries of notches are approximated by circular arcs in hybrid elements.

METHOD

In 1964 Pian introduced the original hybrid method which belongs to the finite element method. After some modifications by Schnack and Wolf, the hybrid method described in this paper was developed by R. Drumm:

The approximation of curved boundaries at notches is executed by a number of polygonal hybrid elements with one circular arc side. The delay behaviour at notches is described by Neuber and can be satisifed by stress interpolation functions of the following terms:

$$f(r,\psi) = ar^{-b}\sin(c\psi) + dr^{-e}\cos(c\psi).$$

This formula can be derived from the Airy stress function satisfying the boundary conditions for circular boundary.

A strict calculus leads to the following stress interpolation functions:

$$\begin{bmatrix} \sigma_r \\ \sigma_\psi \\ \tau_{r\psi} \end{bmatrix} = \lambda_i(\lambda_i-1) \begin{bmatrix} \left[-\left(\frac{r_o}{r}\right)^{-\lambda_i+2} + (2+\lambda_i)\left(\frac{r_o}{r}\right)^{\lambda_i} - (\lambda_i+1)\left(\frac{r_o}{r}\right)^{\lambda_i+2}\right] \cdot \cos\lambda_i\psi \\ \left[-\left(\frac{r_o}{r}\right)^{-\lambda_i+2} + (2-\lambda_i)\left(\frac{r_o}{r}\right)^{\lambda_i} + (\lambda_i+1)\left(\frac{r_o}{r}\right)^{\lambda_i+2}\right] \cdot \cos\lambda_i\psi \\ \left[\left(\frac{r_o}{r}\right)^{-\lambda_i+2} + \lambda_i\left(\frac{r_o}{r}\right)^{\lambda_i} - (\lambda_i+1)\left(\frac{r_o}{r}\right)^{\lambda_i+2} \cdot \right] \sin\lambda_i\psi \end{bmatrix}$$

r,ψ: polar co-ordinates referred to the centre of curvature of the hybrid elements

The λ_i-value can assume every value $\lambda_i \in R$ except $\lambda_i = 0$, $\lambda_i = 1$. The number and the magnitudes of the λ_i-values have to be given by the user of the program system. For appropriate results, magnitudes and numbers of the λ_i-values can be localized.

For crack problems, polygonal hybrid elements are used. The stress interpolation functions applied in this paper contain the asymptotic approximate solution for the short-range field given by Irwin-Sneddon:

$$F_i(r,\psi) = r^{\frac{\lambda_i+4}{2}} \left| M_i \cos\frac{\lambda_i}{2}\psi - N_i \cos\frac{\lambda_i+4}{2}\psi \right|$$

r,ψ: polar co-ordinates referred to the crack-tip

The stress intensity factors can be determined by using special λ_i-values.

The stress interpolation function just described is disposed in a matrix P and the solution of the given problem can be found by determining the matrix β of free coefficients in formula:

$\sigma = P\, \beta$.

FIELD OF APPLICATION

The program system is written for two-dimensional static problems and linear elastic isotropic materials. At the boundary, displacements and point and line loads can be prescribed by the INPUT data.

PROGRAM DESCRIPTION

The hybrid method belongs to the finite element method. In the program system HYBRID, special two-dimensional crack and notch elements are applied in regions of stress concentration, while on the remaining region triangular so-called E2-elements are used.

PROGRAM STRUCTURE

See Fig. 1.

Drawing of the Results

The program PLOT makes a drawing of the generated mesh; the program SPANN draws the boundary stress of a notch.

HYBRID: Determination of Notch and Crack Boundary Stresses

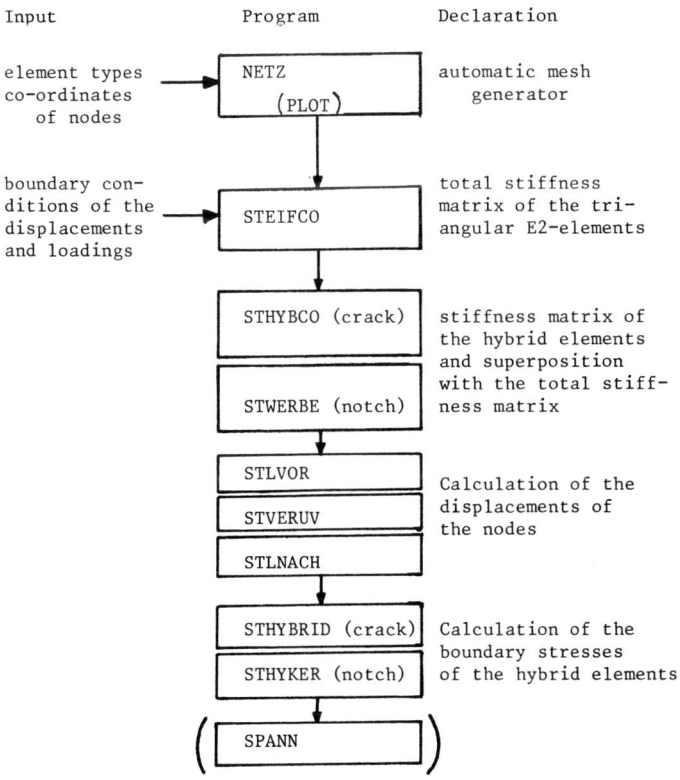

Fig. 1. HYBRID program structure.

Language and Subroutines

The mesh-generator NETZ is written in PASCAL, while all the other main programs and subroutines are written in FORTRAN 77. Subroutines exist for administration of data input and output in the particular programs. Some FORTRAN plot-subroutines are used.

HARDWARE COMPATIBILITIES

The type of the computer is Siemens 7880 with the operating system BS3000. For drawing a BENSON-plotter is used. The program system HYBRID is stored on magnetic tape.

TEST CASE

Problem:
Plate with circular hole under pure shear stress

The stresses in an infinite plate with circular hole are given by analytical solutions of the bi-harmonic differential equation. The loading case of pure shear stresses can be realized by applying loads distributed over a length as shown in Fig. 2. The boundary stress σ_ψ at a circular hole under pure shear stress is given by the formula (1):

$$\sigma_\psi = -4p\cos 2\psi \tag{1}$$

The function σ_ψ can be seen in Fig. 3. Additionally the boundary stress computed by the program HYBRID is illustrated in Fig. 3.

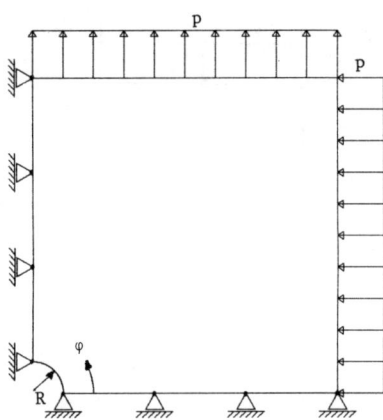

Fig. 2. Symmetrical problem: boundary conditions.

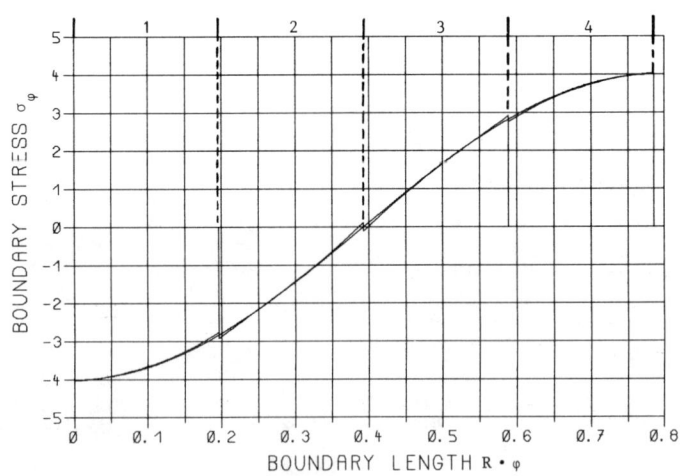

Fig. 3. Boundary stress of the circular hole under pure shear stress.

For this test case, four hybrid elements are disposed as shown in Fig. 4.

The remaining part of the plate is simulated by two-dimensional triangular elements with six nodes each. In this example 46 elements were used for the plate with a dimension size of 320*320. The diameter of the circular hole is 1.

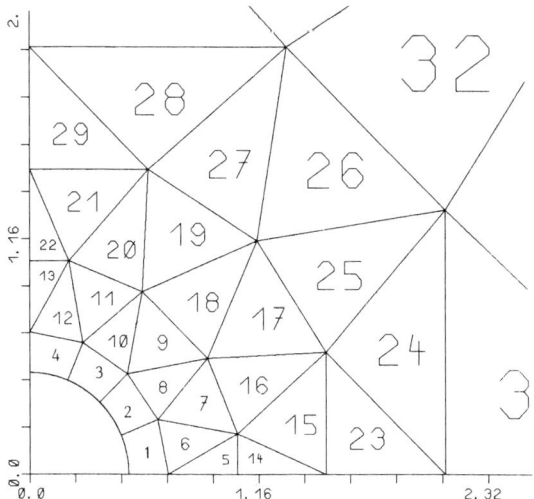

Fig. 4. Location of the hybrid elements.

The total number of nodes is 119, which implies 238 degrees of freedom. The bandwidth of the total stiffness matrix is 21.

INPUT-Data:

- co-ordinates of the nodes
- boundary conditions for the loadings and displacements as shown in Fig. 2
- λ_i-values for the stress interpolation functions:

i=	1	2	3	5	6	7	8
λ_i	0.05	2.0	4.5	11.0	15.0	19.5	24.5

OUTPUT-Data:

- new co-ordinates and numbers of nodes
- new node numbers for boundary conditions
- λ_i-values of the stress interpolation functions
- displacements of all nodes
- boundary stresses of the hybrid elements

OUTPUT-Plot

- generated mesh (Fig. 4)
- boundary stress distribution (Fig. 3) respected and calculated

Computation time: 40 sec.

References: Industrieanlagen Betriebsgesellschaft m.b.H., (IABG), Munich, where the program system is called AVA (Advanced Variational Principle Analysis)

PRACTICAL EXAMPLE IN INDUSTRY

In industrial processing engineering drums were used for mixing fluids, as shown in Fig. 5. The firm IABG, which have implemented the HYBRID-system in the program system NASTRAN, has examined the bottom A of such a mixing machine. Of special interest were the stress concentrations at the occurring cracks at the lines of weld.

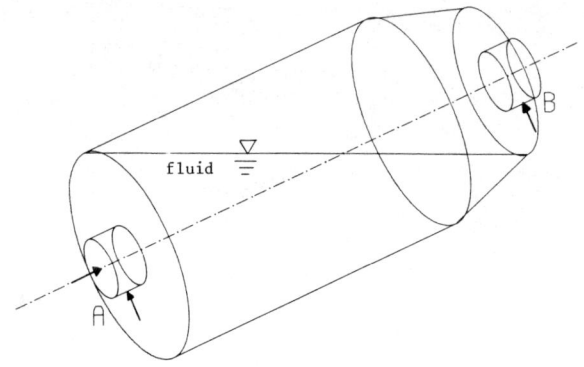

Fig. 5. View of a mixing drum.

The examined body is an axi-symmetric double-drum botton under non-axisymmetric loading.

The discretization was realized by dividing the body in 96 geometric identical segments, each coating an angle of $3.75°$ degrees (see Fig. 6).

Isoparametric elements were used. For each segment 370 nodes resulted, which correspond to 1010 degrees of freedom. Additional special crack hybrid elements were used at different places.

To diminish the costs the weight of insertments and fluid was added to the own weight. The calculation was executed for a bottom of 8 mm of thickness. For comparison the thickness was varied to 5 mm. It was found, that the small thickness is also suitable for the construction, because of the comparable low reference stresses the lines of weld, however, are then loaded essentially heavier by fulling.

INPUT-DATA

- co-ordinates of the nodes for mesh generation
- boundary conditions for loadings and displacements
- λ_i-values for the crack elements

OUTPUT-DATA

- displacement vector
- boundary stresses of hybrid elements
- stress intensity factors

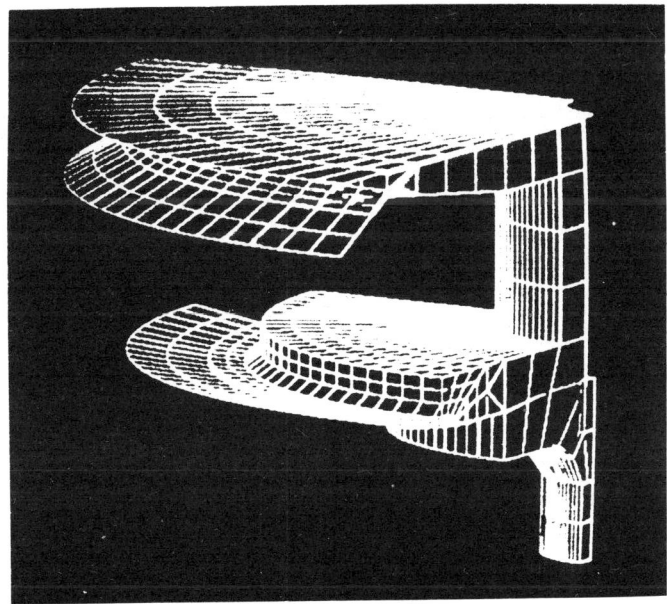

Fig. 6. Screen drawing of a drum-bottom (segment).

OUTPUT-PLOT

- generated mesh (Fig. 6)
- segments under displacements

Computation Time

48 sec.

Facilities

Several computers are available for the NASTRAN-program-system (e.g. Siemens 7880).

This problem was computed on a CYBER 175-system.

INFESA: INTERACTIVE FINITE ELEMENT STRESS ANALYSIS

A. O. Moscardini and B. A. Lewis

Department of Mathematics and Computer Studies, Sunderland Polytechnic, Sunderland, UK

ABSTRACT

INFESA is an interactive software package principally dealing with two dimensional stress-strain analysis. It is written in BASIC and, at present, runs on a VAX 11-750 with GINO graphics. The package is fully interactive in the sense that it is self explanatory (with help available at every stage) and also that all datafiles are prepared and processed automatically. There are four major stages in the analysis, all of which used independently:

(a) the generation of the element mesh,

(b) the formulation of the specific problem,

(c) the solution of the problem,

(d) the presentation of results.

The mesh is constructed with the minimum data input by the user and three, four, six or eight noded elements can be produced.

Boundary conditions, material properties and the nodal force distribution are automatically calculated in response to clear direct questions.

The modular structure allows the easy substitution of other solvers allowing the package to be used for different problems. A solver called HECFE has been written and used successfully for heat conduction problems.

The plethora of results that such problems produce can be interrogated by the user and presented in understandable form. Graphics are also available at each stage.

THEORETICAL BACKGROUND

The growth of versatile finite element packages has been at the expense of more complicated software engineering and sophisticated hardware requirements. There is still a need for a user oriented interactive package that will run on a small machine with little need for extra hardware. INFESA, Interactive Finite Element Stress Analysis, is such a system which has been designed over the last 5 years.

At the planning stage, the user was quite clearly identified. No extensive

computing knowledge is needed. Obviously some knowledge of the use of Finite
Elements in solution of stress strain problems in two dimensions is needed but the
software is intended for an undergraduate who is meeting such theory for the first
time and is further intended to complement this learning in the sense that it is
a teaching aid. It must also allow for the post graduate who wishes to solve real
problems and so different levels of usage must be available.

The software was designed to be portable in the sense that it will run efficiently
on any minicomputer that supports BASIC. It will not therefore, rely on any
machine dependent functions or special BASIC commands.

The following developmental objectives were also proposed:

1. The software is interactive.
2. Information is expected in a form convenient to the user, not, as is more
   common, convenient to the program.
3. Data preparation for the user is kept to a minimum.
4. Plotting routines are available in the program independent of the computer on
   which it is mounted.

After careful consideration, the data structure was organized in an hierarchical
fashion with easily replaceable independent modules.

Although this method does not utilize hardware and software capabilities to the
full, it is simple and the construction of output files at various key stages
enables the student to use the software in a very flexible manner. Care was taken
to prevent the package from being I/O bound and the number of output files has
been reduced to only eight.

The internal structure of INFESA is shown in Fig. 1. It is structured in an
hierarchical manner into five levels. The four principal stages occur at level
two which are:

(a) the generation of the element mesh   FEMGEN

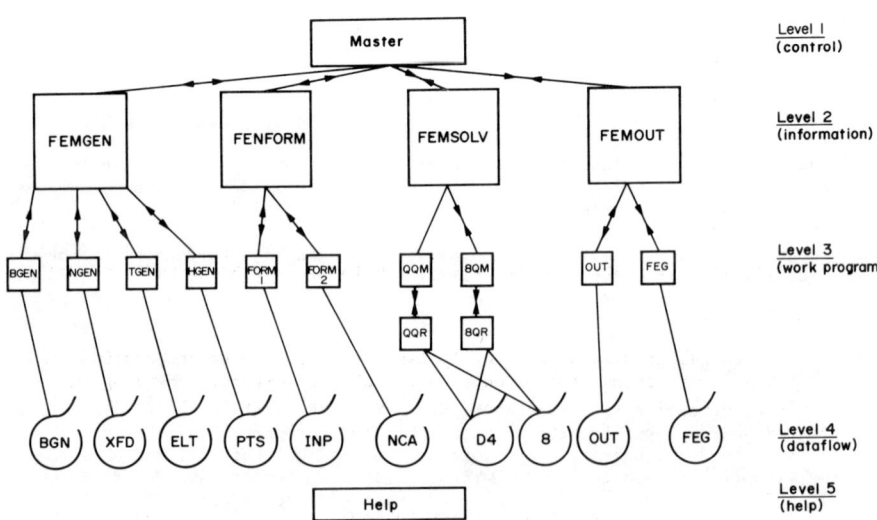

Fig. 1.

(b) the formulation of the finite element problem   FEMFORM
(c) the numerical solution to the problem   FEMSOLV
(d) the interpretation of results   FEMOUT

Each stage is self contained, complete and can be run independently. Suitable data files are generated at the appropriate time and clear instructions are given to the user on how to proceed. A top-down approach was followed in the overall design but each individual module used a bottom up approach. This seems to be a very flexible structure. For example, other applications of the finite element method could be dealt with by the modification of the FEMFORM stage.

The appearance of the software to the user and the design and structure of the user interface is covered by objectives one to four.

The language was chosen to be standard ANSI BASIC, and at present runs on a VAX 11-750 with GINO graphics as an extra option.

Thus the software does not rely on any specialized hardware or software but uses any facilities that are available. A change to another machine or an upgrade in the present graphic facilities would only entail the substitution of one subroutine in each module.

The structure is flexible in the sense that the user can stop and start at will, by preserving the requisite data files. The MASTER program, is used as a substitute for user documentation. Help is available in every section but there is also a level five program called HELP. This contains details of each module and each data file. As the program HELP presupposes that the names of the individual files are known, it is expected that this will be more beneficial to the more confident user than the complete novice who can use the software without any knowledge of the data files.

FIELD OF APPLICATION

Geometrical two-dimensional.

Linear elastic materials with composite materials.

Analysis capabilities - static.

Loadings - pressureloads, point loads, gravity loads, temperature distributions.

PROGRAM DESCRIPTION

Finite elements

Two-dimensional solid isoparametric elements with 3, 4, 6 and 8 noded elements.

Program structure.

The program is designed to be completely user friendly. Data preparation and input is kept to a minimum and is checked for obvious errors. The program is interactive in the sense that the user has the opportunity to change every item of data input and the format of this input is fully explained. The graphics are automatically called whenever possible.

A complete, automatic mesh generator, FEMGEN, is an integral part of the program. FEMGEN produces a boundary definition, an internal nodal mesh, triangulation and/or quadrilation of the region. The nodes are renumbered for optimum bandwidth according to the Cuthill-McKee algorithm but all user instructions and graphics refer to the original numbering except where renumbering is expressly demanded.

There is also a postprocessor FEMOUT, which sorts through the plethora of results and presents the information in an understandable form which includes pictures of the transformed or distored region and contour plots of stresses and strains.

The language is standard BASIC and the overall structure is shown in Fig. 1. The modular structure allows for new subroutines or modifications to be performed very easily thus making the program extremely portable.

HARDWARE COMPATABILITIES

VAX 11-750 with 1 Mb using VAX/VMS V3.1.

VAX II BASIC (for interacting debugging) TECTRONS 4010 compatable terminal.

VAX 11-750

In house developed graphics conforming to TECTRONICS 4010.

VAX/VMS

Magnetic tape.

EXAMPLES OF APPLICATIONS

Test Case. The following problem was chosen as a test case as exact analytic results are available.

The cantilever problem. Consider a steel cantilever beam subject to an end load of 35 KN/mm^2 Youngs modulus is taken as 210 KN/mm^2. Poissons ratio is 0.25 and the moment of inertia is 8/3 x 10^2 mm^4 (Fig. 2.)

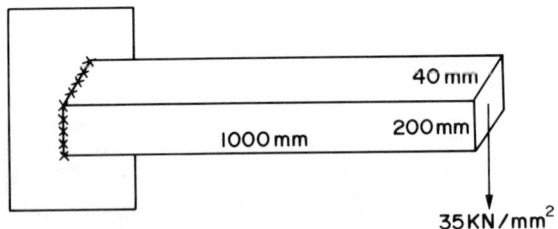

Fig. 2.

Analytic results (see reference 1):

$$V_{(y=1000)} = \frac{Px^3}{6EI} - \frac{P\ell^2 x}{2EI} + \frac{P\ell^3}{3EI} = 2.0833 \text{ mm at free end}$$

There is an extra shearing force of $\frac{P\ell^2(\ell-x)}{2IG}$ = .078125 mm

i.e. Max deflection of 2.16 mm at free end

$$\sigma x = \frac{-Pxy}{I} \quad \text{i.e. max stress = 131.25 at fixed end}$$

INFESA: Interactive Finite Element Stress Analysis

also at y = 0

$$\tau xy = \frac{-P}{2I} ({}^2-y^2) \text{ i.e. } -6.5525 \text{ at free end}$$

The beam was discretized into 4 eight noded isoparametric elements as shown in Fig. 3.

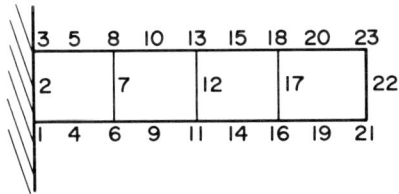

Fig. 3.

The results show displacements and stresses. It can be seen that the displacements at the furthest end from the wall agree to one decimal place with the analytic results as do the maximum stresses in the x direction. The slight variation in the other stresses can be explained by the discretization.

Displacements For Lead Case Load

Node	Displacements
1	-0.00000000, -0.00000000
2	-0.00000000, -0.00000000
3	-0.00000000, -0.00000000
4	-0.06990728, -0.05354839
5	-0.06990728, -0.05354839
6	-0.13314474, -0.19031225
7	-0.00000000, -0.18235880
8	-0.13314474, -0.19031225
9	-0.18774919, -0.39694815
10	0.18774919, -0.39694815
11	-0.23175743, -0.66521698
12	-0.23175743, -0.66521698
13	0.23175743, -0.66521698
14	-0.26569801, -0.98322129
15	0.26569801, -0.98322129
16	-0.29009949, -1.33768938
17	-0.00000000, -1.33557517
18	0.29009949, -1.33768938
19	-0.30481085, -1.71631230
20	0.30481885, -1.71631230
21	-0.30968108, -2.10717371
22	-0.00000000, -2.10725425
23	0.30968108, -2.10717371

Nodal Average Stresses

Node	σxx	σyy	τxy
	-131.250	-32.812	-4.375
2	-0.000	-0.000	-4.375
3	131.250	32.813	-4.375
4	-114.844	-12.009	-4.375
5	114.844	12.009	-4.375
6	-98.438	8.795	-4.375
7	-0.000	-0.000	-4.375
8	98.438	-8.795	-4.375
9	-82.031	3.214	-4.375
10	82.031	-3.214	-4.375
11	-65.625	-2.368	-4.375
12	-0.000	0.000	-4.375
13	65.625	2.368	-4.375
14	-49.219	-0.846	-4.375
15	49.219	0.846	-4.375
16	-32.813	0.677	-4.375
17	-0.000	-0.000	-4.375
18	32.813	-0.677	-4.375
19	-16.406	0.169	-4.375
20	16.406	-0.169	-4.375
21	-0.000	-0.338	-4.375
22	-0.000	-0.000	-4.375
23	-0.000	-0.338	-4.375

Practical Examples in Industry

Firing of silica shapes. The manufacture of silica shapes for coke oven construction has always been difficult with cracking a common cause of failure. The raw shape needed for firing, called the "green shape", is produced by placing silica material into a mould and subjecting it to high pressures of two to three tonnes p.s.i. If frictional forces at the wall are taken into account, then the boundary conditions are nonlinear and the problem becomes a difficult one. The mathematical problem is thus to determine the stress-strain relationships developed in a green silica shape, when a constant force operates on the upper boundary and nonlinear boundary conditions apply at the walls.

Application area. Structures.

Type of problem. Stress-strain with nonlinear boundary conditions.

Iterative scheme.

1. Set $S = S_1$ (arbitrary slip point).

2. Replace the boundary condition $\tau_{xy} = \mu\sigma_x$ by

$$\tau_{xy} = \mu\sigma_x = -\frac{\mu\nu p}{1-\nu}$$

which would be true if we had a smooth boundary (μ is the coefficient of friction, ν is Poisson's Ratio).

3. Solve, using the linear finite element method. This gives values for τ_{xy} on $x = a$, $0 \leqslant y \leqslant S$. Guess a new value for S, S_2 where if

$$\frac{\tau_{xy}}{\sigma_x} \begin{cases} > \mu \text{ at } x=a, y=S- \text{ then } S_2 < S_1 \\ < \mu \text{ at } x=a, y=S+ \text{ then } S_2 > S_1 \end{cases}$$

4. Iterate the above scheme until

$$\left|\frac{\tau_{xy}}{\sigma_x} - \mu\right| \leqslant \varepsilon \text{ and } S = \bar{S}$$

5. Determine σ_x on $x=a$ and reset $\bar{\sigma_x}$ to these values (NB $\bar{\sigma_x}$ is now a function, not a constant).

6. Repeat steps 3 to 6 until S and $\bar{\sigma_x}$ converge.

Diagram (Fig. 4).

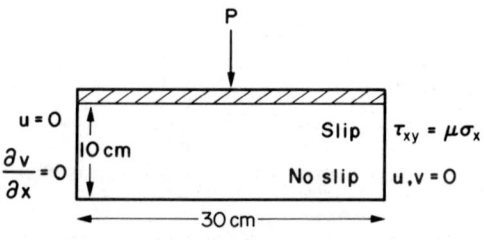

Fig. 4.

Discretization. Rectangular mesh.

Type of element. There were 176 nodes forming 150 four noded isoparametric quadilateral elements.

Degrees of freedom. Each node has two degrees of freedom.

INFESA: Interactive Finite Element Stress Analysis

Bandwidth. As a variable bandwidth is used, this is not applicable.

Part of program used. The whole program was used. The mesh was generated by FEMGEN the problem formulated in FEMFORM, solved in FEMSOLV and results were produced by FEMOUT. It was run on a VAX 11-750 with Gino Graphics.

Input and output data. The input data can be split into three sections:

(a) geometrical — for generation of mesh

(b) material properties — values of constant force, Youngs Modulus, Poissons Ratio

(c) boundary conditions

The output data was concerned with the location of the slip point and then the production of a contour map of the stresses and strains.

Computation time. This is of the order of 1 min c.p.u. on VAX 11-750.

Shelf-angle flooring. This is an I shaped column, partly protected by a brick or concrete wall which is subjected to a British Standard fire. This involves a large thermal shock to the structure and the ambient temperature is thought to obey the relationships

$$T = T_0 + 345 \log_{10} (8t+1)$$

Type of problem. Heat conduction in solids.

Diagram (Fig. 5).

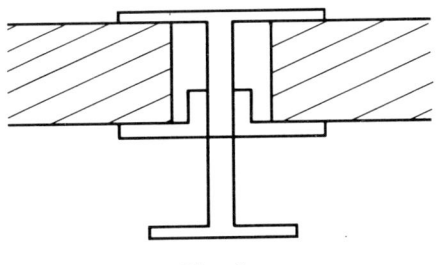

Fig. 5.

Discretization. Three noded triangular elements.

Type of elements. There were 104 nodes forming 105 three noded triangular elements.

Degree of freedom. Each node has two degrees of freedom.

Bandwidth. As a variable bandwidth was used, this is not applicable.

Part of program used. The whole program was used. The mesh was generated by FEMGEN, the problem formulated by FEMFORM, solved by FEMSOLV - using HECFE and output generated by FEMOUT. This problem was run on a DEC PDP 11-40.

Input and output data. The input data can be categorized into:

geometrical data - for generation of mesh

material properties.

The output file consists of temperature distribution.

Computation time. This is of the order of 2 min.

REFERENCE

1. Timoshenko, S. and J. N. Goodier (1951). *Theory of Elasticity.* McGraw-Hill, New York, pp. 51-56.

MEF/MOSAIC: LINEAR, NONLINEAR FINITE ELEMENT CODE WITH INTERACTIVE GRAPHIC DISPLAY

J. F. Cochet

CSI/UTC, Centre de Recherches de Royallieu, B.P. 233, 60206 Compiègne, France

ABSTRACT

MEF is a finite element code to solve 2 or 3D problems of solid mechanics, fluid mechanics, heat transfer, electromagnetism, fluid structure coupling, piezoelectricity and others.

MEF was born in the nearly 80s. It is modular, easy to implement on microcomputers and includes interactive tools for graphic display (pre-post processor MOSAIC).

MEF is written in standard FORTRAN with precise programming norms and is easy to implement, modify, develop and to maintain. The program has been implemented on various 32-bit computers. MEF has been jointly developed by professors, engineers, graduate students from the University of Technology of Compiègne and from the University Laval of Quebec City.

MEF is industrialized and sold by the company C.S.I. (Compiègne Science Industrie).

Its purposes are to be used:

(a) as a very adaptative tool by companies which already own finite element codes.

(b) as teaching and research tool by institutes, universities and research centers.

(c) by companies which need specific developments in their fields.

THEORETICAL BACKGROUND

The theoretical background and the architecture of the program is described in detail in reference 1. To sum up, the titles of the chapters are:

. approximations with Finite Elements

. various type of elements

. variational formulation of engineering problems

. numerical procedures

. coding techniques

This chapter describes the overall organization, the database organization, a description of the functional blocks, and applications.

FIELDS OF APPLICATION

Geometrical. 2D, axisymmetric, 3D; beams, plates, shells.

Materials. Elasticity linear, nonlinear; elasto-plasticity; viscoplasticity; hyperelasticity.

Analysis capabilities. Static, structural vibrations, dynamic, geometrical nonlinear.

Contact.

Buckling, post-buckling.

Fracture mechanics: stress intensity factors determination (K, J, G); 2D crack propagation.

Shape, mass, stress optimizations.

Thermal analysis.

Free surface fluid flow, 3D incompressible fluid model, flow through porous media.

Loadings. Concentrated loads, imposed displacements, uniformly distributed loads, pressure, gravity, centrifugal, residual stresses, thermal stresses, temperature.

PROGRAM DESCRIPTION

Method

MEF uses the displacement finite element method

Type of Elements

The MEF element library contains:

- 3 beam elements
- 4 2D elasticity
- 4 axisymmetric elasticity
- 6 3D elasticity
- 10 2D/3D quasi-harmonic
- 2 plates (DKT,DKQ)
- 4 shells (thin, thick)
- specific elements: pressure, connecting elements, contact

Program Structure

MEF contains:

- a kernel of about 10,000 FORTRAN orders for the typical functions of the FE method used for all the problems

  - input data

- assembly process
- sky line storage of the matrix
- solver
- data generation for postprocessing

This kernel has been described in detail in references 1 and 2.

- an element library which can be enriched or modified to model trusses, beams, plates, shells, continuous media, fluid media, piezoelectricity equations. These elements solve different kinds of problems:

  - linear and nonlinear
  - stationary and nonstationary
  - static and dynamic
  - eigenvalues

- standard execution blocks:

  - in-core linear solver
  - out of core linear solver
  - nonlinear solver
  - nonstationary solver
  - eigenvalues extraction

- specific blocks and elements

  - buckling eigenvalues and vectors
  - shape optimization
  - frictional contact
  - post-buckling analysis
  - plastic or viscoplastic behaviour
  - large deformations
  - non self similar crack propagation
  - incompressible fluid flow
  - free surface fluid flow (rivers, ...)
  - porous media fluid flow
  - heat transfer
  - ...

The data input can be done as usual step by step following the user's manual (free format facility) or by using the preprocessor MOSAIC.

MOSAIC is an interactive graphic pre-postprocessor to generate input data for MEF. The geometry of the components is defined through points, lines, surfaces, and volumes to be meshed. The meshes capabilities are given in Table 1.

TABLE 1

type	method
2D	reference surface
2D	meshed boundary
2.5D	shells
3D	reference surface plus bricking
3D	meshed boundary surface

Automatic renumbering and bandwidth minimization are included.

The postprocessing from the results of computation (deformed shape, stresses, ...) is interactively carried out. Graphic display is done for isolines, isocolour plots, deformed shapes and hardcopy are available through the terminal or through a plotter.

The subroutine structure of MEF allows fast modifications and is as follows:

- Main program

    The user controls the execution of the different functional blocks through the main program by calling the proper subroutines. The main program is made of two parts (Fig. 1).

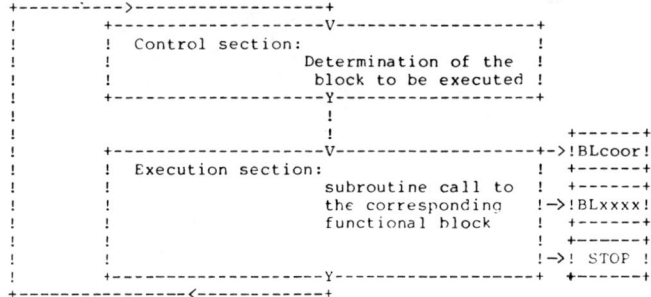

Fig. 1.

- Input data blocks:

    These blocks read the data used for the modelization of the physical problem: nodes coordinates, connectivities, element properties, boundaries conditions, loads, ...

    The data generated by MOSAIC are readable directly without any extra manipulation.

- Execution blocks:

    All the different used blocks have a similar structure. They all build up elementary matrix and vectors, global matrix and vectors by assembling the elementary ones, solve a linear equations systems, print and/or store the results.

    The differences are only in the manner these operations are linked together. The global structure of these blocks can be summarized as in Fig. 2.

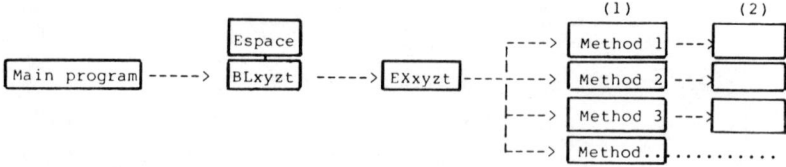

Fig. 2. (1) Choice of the solver method, (2) Common subroutines for the assembling process, the factorization, the printing and results file generation.

MEF/MOSAIC: Linear, Nonlinear Finite Element Code   99

HARDWARE CAPABILITIES

- 32-bit processors
- Virtual memory (useful for large arrays)
- Memory
    . in-core memory 3Mo   (MEF + MOSAIC + Data)
    . mass storage 20-30Mo (MEF + MOSAIC + Data)
- Peripherals
    . 1 graphic display terminal
    . 1 visual display terminal
    . 1 plotter (optional)
- Supported computers
    . DEC VAX operating system VMS
    . PRIME operating system PRIMOS
    . UNIVAC 1100
    . IBM 4341/3081
    . BULL MINI 6 (MEF only)
- Supported terminals (MOSAIC)
    . TEKTRONIX 40XX
    . TEKTRONIX 41XX
    . ENVISION 215/220/230
    . WESTWARD 2014/2119/2115/3220
    . RAMTEK   62XX
    . All Compatible TEKTRONIX 40XX
- Media
    Magnetic tape 1 600 bpi
        Backup format (VAX/VMS)
        Fixed format 80 characters (EBDCIC or ASCII)

EXAMPLES OF APPLICATIONS

Threaded Joint for Casing and Tubing (Figs. 3, 4, 5)

Description of the components. The connecting joint is made of a threaded part which connects the two pipes. A conical thread and two tapered-seals and shoulders keep the joint leakproof.

Application area. Direct application area: fuel extraction fields. Indirect application area: extension to all connectors with thread.

Type of problem. The aim of the modelisation was to achieve the behaviour in elasticity and plasticity of the connector.

Elements axisymmetric quadratic triangle and quadrilateral

6 nodes contact element

Number of elements	: 604
Number of nodes	: 2070
Number of degrees of freedom:	4140
Maximum bandwidth	: 187
Average bandwidth	: 64

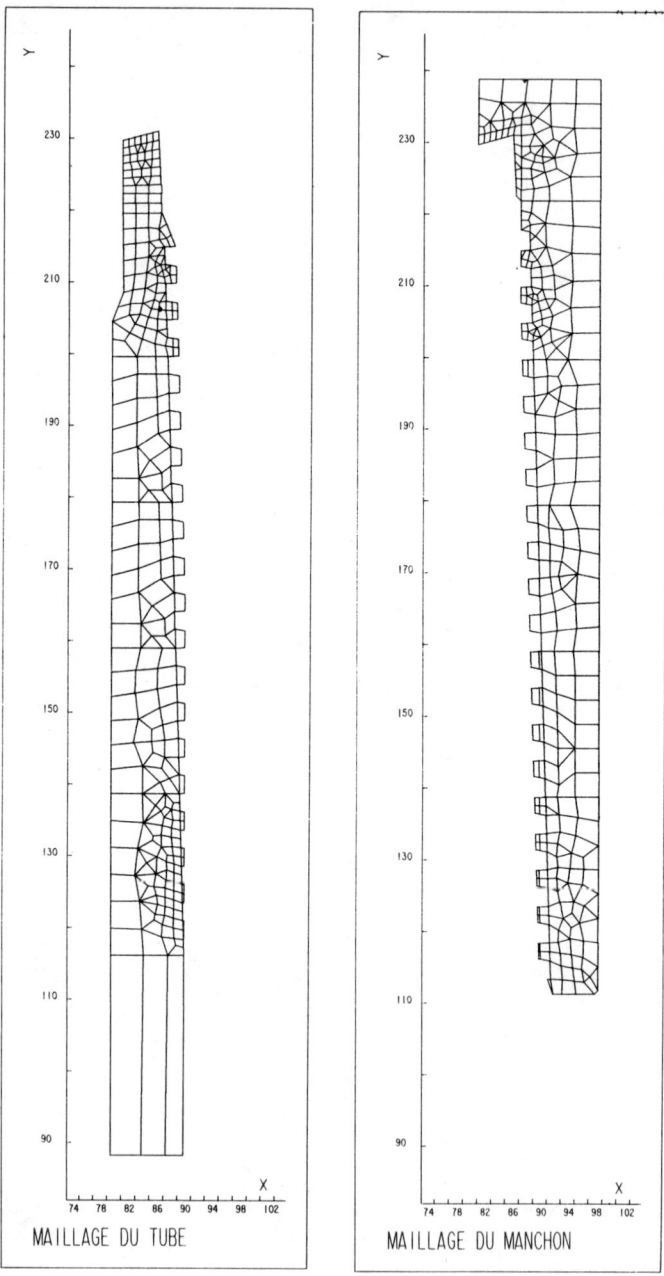

Fig. 3. Initial mesh.

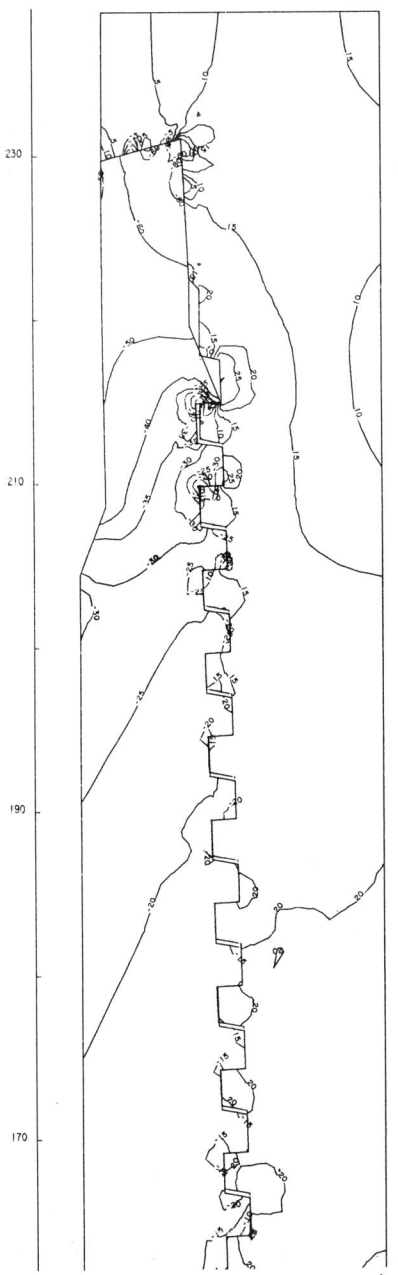

Fig. 4. Stress isovalues plot.

Computer   : VAX 11/780
Peripherals   : Tektronix 4105
                     Hardcopy
                     Benson plotter
Computing time   : 15 minutes CPU for each interaction (computing time between 1 and 4 hours depending on the type of model)

Shock Absorber (Figs. 6, 7)

Description of the part. Part of a bumper.

Application area. Large displacements shells behaviour (moderate rotation).

Type of problem. The aim was to attain the displacement reached under large displacement (quasistatic study)

Elements triangular thin shells
(18 degrees of freedom)

Number of elements: 1061

Number of nodes   : 534

Maximum bandwidth : no statistics available

Computer   : VAX 11 780

Peripherals   : Tektronix 4105
                     Hardcopy 4695
                     Benson Plotter

Computing time   : no statistics available

Disk Rotor Shape Optimization (Fig. 8.)

Description of the components. Turbine blades are hold on rotor disks. The centrifugal loads due to rotation create very large stresses where the geometry is complex: A crack can nucleate where the stresses concentrate.

Application area. The application area is mainly where it encounters shape optimization problems with multiple constraints which have to be numerically solved due to the large number of constraints and unknowns.

Type of problem. The objectives are to minimize the stress concentration in the rotor disk due to centrifugal force on the blade including the sliding contact between the disk and the blade.

Elements   : quadratic triangles and quadrilaterals
Number of nodes   : 419
Number of elements   : 134
Number of degrees of freedom   : 858
Maximum bandwidth   : no statistic available
Average bandwidth   : no statistic available
Computer   : VAX 11/780
Peripherals   : Ramtek 6212 + Benson plotter
Computing time   : 10 minutes CPU for 5 iterations

Fig. 5. Isovalues of radial stresses.

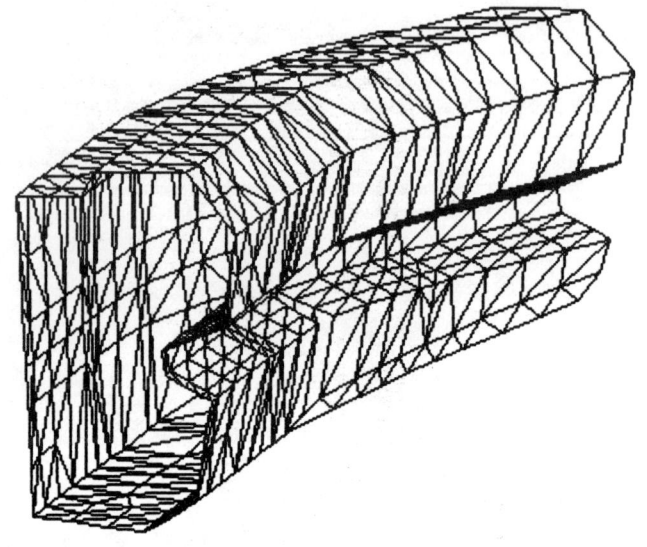

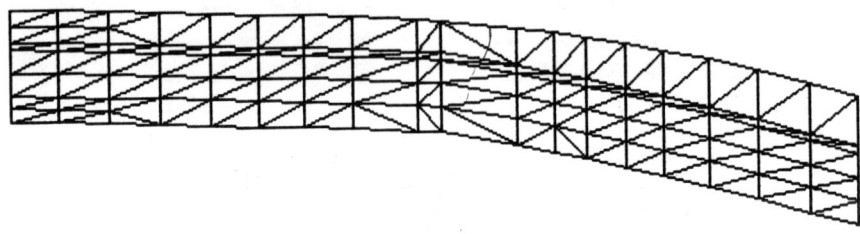

Fig. 6. Initial mesh.

St Laurent River (Figs. 9-11)

Description of the part. A factory implantation on a river sometimes needs to know the chemical or thermal wastes or rejects impact.

Application area. Model of free surface fluid flows in rivers, estuaries, bays and environment diffusion problems.

Type of problem. The aim is to model the flow in the St Laurent River at different rates and to compare the influence on the rejection zone.

Elements	: 2D quadratic triangles (degrees of freedom, u, v, h) with integration of speed in the vertical.
Number of nodes	: 809
Number of elements	: 374
Number of degrees of freedom	: 1614
Average bandwidth	: 90

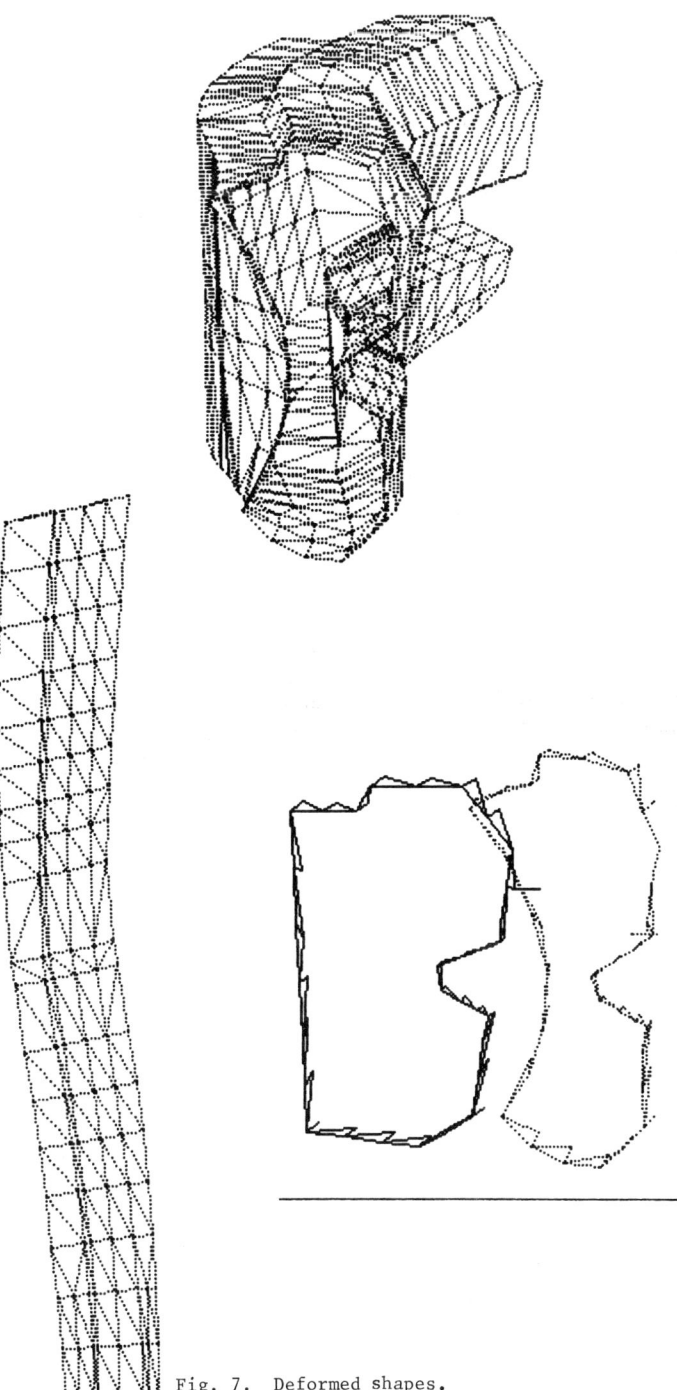

Fig. 7. Deformed shapes.

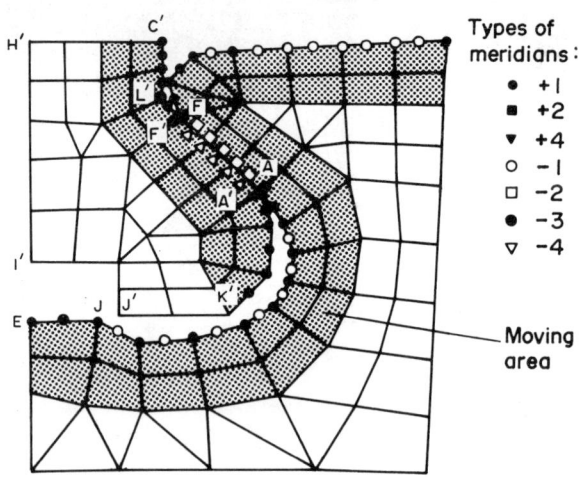

Fig. 8ii. Initial mesh.

Computer   : DEC VAX 11/780
Peripherals   : Benson plotter
Computing time   : 90 minutes CPU for 10 iterations

Pressure Vessel (Figs. 12, 13)

Description of the part. Cylindrical pressure vessel under internal pressure.

Application area. This type of component is mainly used in nuclear power generators.

MEF/MOSAIC: Linear, Nonlinear Finite Element Code

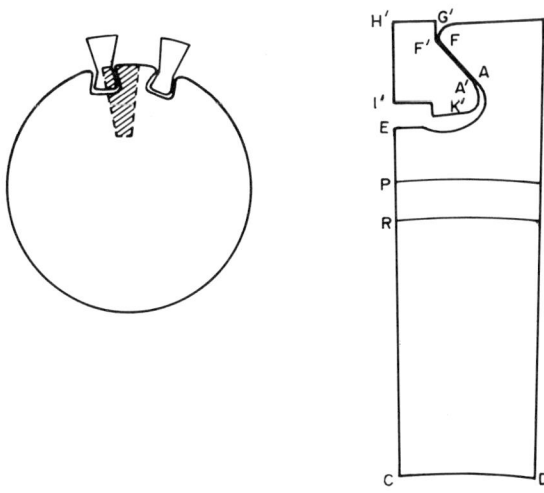

Fig. 8i. Problem description.

Type of problem. Buckling eigenvalue extraction. Nonlinear analysis with elastoplastic material behaviour.

Elements triangular thin shells

Number of elements	: 480
Number of nodes	: 84
Number of degrees of freedom	: 504
Computer	: VAX 11/780
Peripherals	: VT/100 + benson plotter
Computing time	: no statistics available

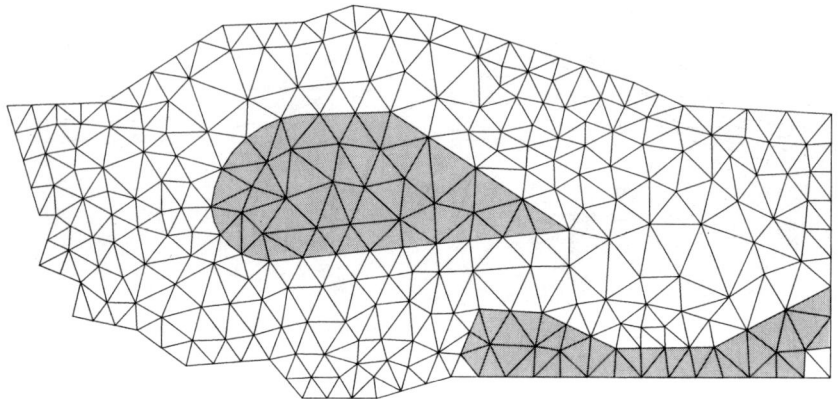

Fig. 9. Mesh.

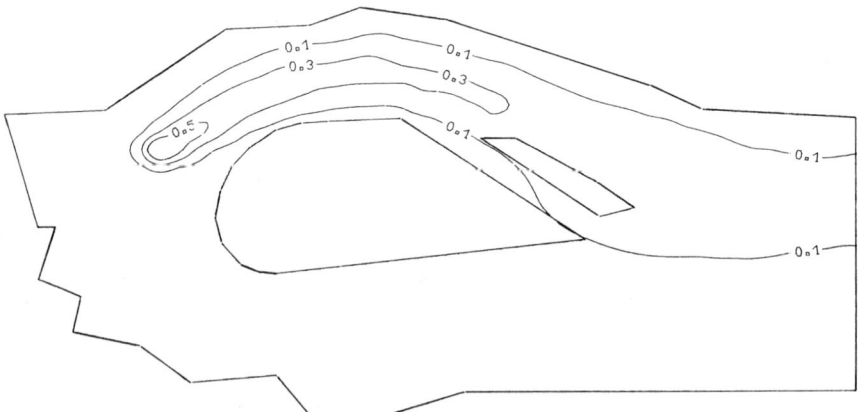

Fig. 10. Thermal discharge.

MEF/MOSAIC: Linear, Nonlinear Finite Element Code 109

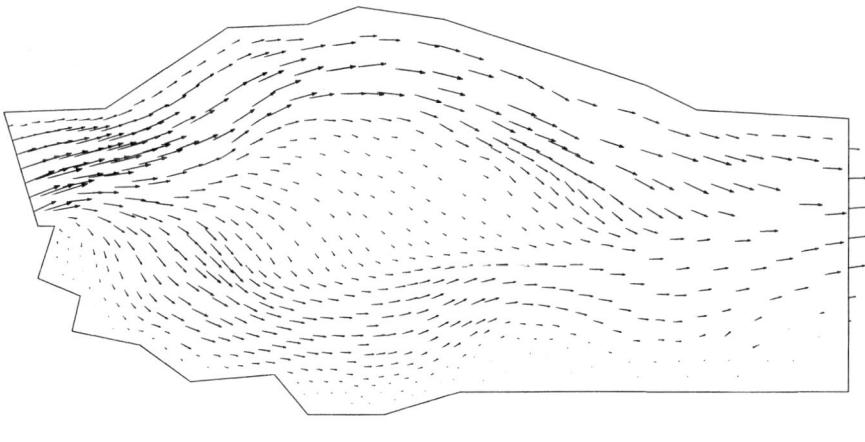

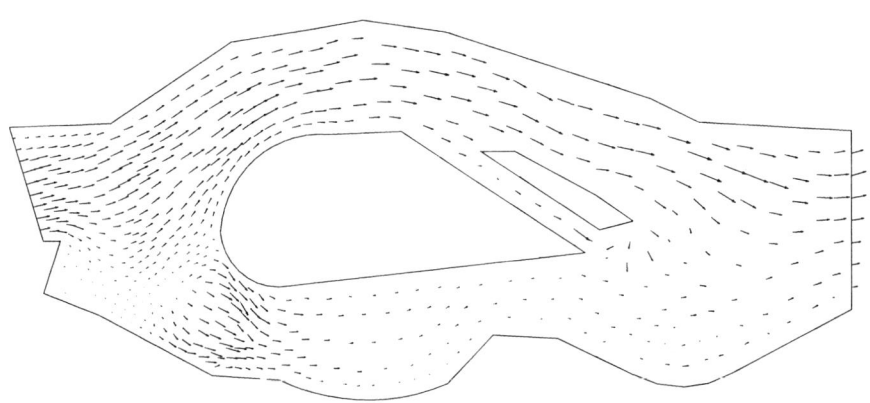

Fig. 11. Velocity profile.

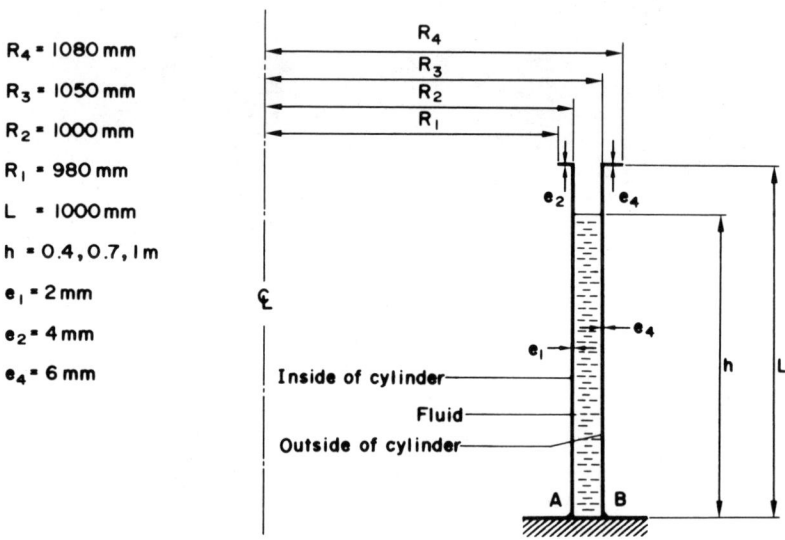

Fig. 12. Problem type.

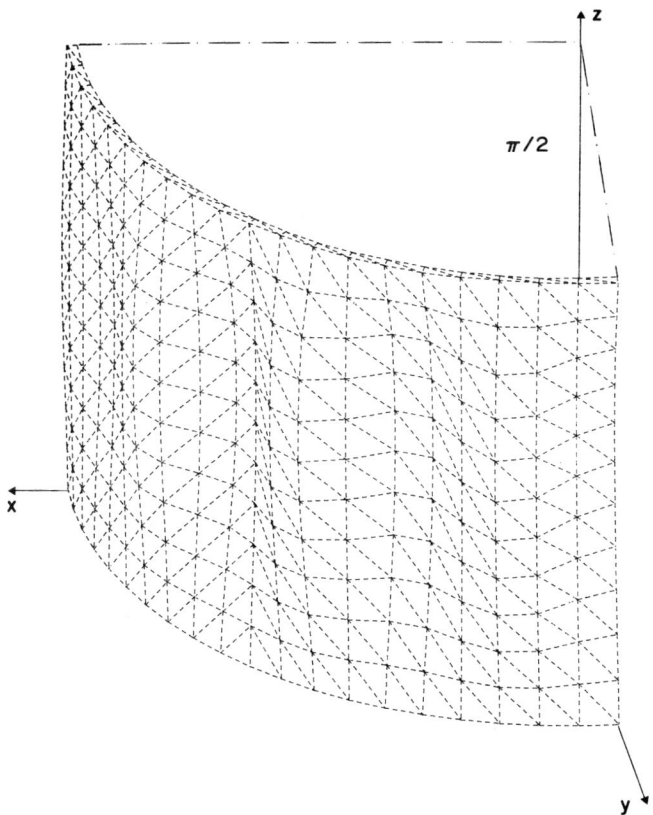

Fig. 13. Deformed shape.

REFERENCES

1. Dhatt, G. and G. Touzot (1981). *Une Présentation de la Méthode des Eléments Finis*. Editions Maloine.

2. Dhatt, G. and G. Touzot (1984). *The Finite Element Method Displayed*. Wiley.

RCAFAG: ON THE STABILITY ANALYSIS OF REINFORCED CONCRETE ARCHES AND FRAMES IN THE GEOMETRICAL AND MATERIAL NONLINEARITIES FIELD

L. Cappellari and R. Gori

Istituto di Scienza delle costruzioni, Università di Padova, Italy

ABSTRACT

RCAFAG is a dual finite element programme for structural analysis based on the flexibility method.

This is an enhanced version of AFAG, designed particularly for the solution of reinforced concrete structures, but also suitable, more in general, for solving structures in which the nonlinear behaviour of the materials must be taken into account as well as the problems connected to geometrical nonlinearities.

RCAFAG permits the static, linear and nonlinear structural analysis of reinforced concrete structures such as arches and similar r.c. structures (bridge arches, portal frames, tubes, etc.).

The input is extremely articulated and allows for the implementation of diverse external loads and pre-strain conditions as cause of stress (shrinkage, creep, thermal variations, inelastic constraint displacements, etc.).

Prints and/or graphics of the internal forces of single loading and of the relative deformed shapes can be requested in output.

THEORETICAL BACKGROUND

The method is iterative, converging at the only congruent solution. During each step of the iteration the geometry and stiffness of the structure are considered constant (linearization), giving, at the end, new displacements and stiffnesses.

The flexibility method has been used to give a drastic reduction in the number of unknowns, as in AFAG.

The structure is divided into as many elements as desired. The inertia moment of the section is constant or varies in a linear fashion along each element. Each element is short enough to be ideally equal to a "section" of infinitesimal length, and has the geometrical, mechanical and material properties of such a section. Before performing the analysis, the new stiffnesses are assigned for each element during each iteration. These are determined from the stresses of the previous iteration and the laws of the materials.

Each step of the iteration is performed considering the structure in the deformed shape and with an elastic linear behaviour. The displacements are determined later using the virtual force theorem.

The flexibility coefficients of the elasticity equations are obtained, element by element, by integrating the products of linear or parabolic force functions along beam elements with linearly variable cross sections.

The knowns are obtained in the same way.

The same algorithms used for the determination of the unitary equilibrated states on the statically determined systems can be used for the evaluation of the displacements u, v and φ by means of the virtual force theorem.

Stress-strain law $\sigma - \varepsilon$ for concrete and steel was assumed, as in the FIP-CEB model code 10.4.3 and 10.4.4, although any suitable law may be assumed.

The behaviour of reinforced concrete is represented by the relations between moment and curvature and between normal force and normal strain at the barycentric axis of the arch.

Since there is no linear relation between the above-mentioned magnitudes, an iterative procedure has been developed for the determination of the ε normal strains at the upper and lower edges of each section given known values of $\overline{M}$ and $\overline{N}$, of the geometrical and mechanical characteristics of the sections and of both types of material. A brief illustration of the procedure is given below.

$\overline{M}$ and $\overline{N}$ are known; the expressions of the $M(\varepsilon_1,\varepsilon_2)$, $N(\varepsilon_1,\varepsilon_2)$ parameters, functions respectively of the normal strains at the lower and upper edges, are found from the $\sigma - \varepsilon$ diagrams of the materials and from the characteristics of the section by integration of the elementary moments of the tensions in the whole section. The values of ε_1 and ε_2 are obtained by the solution of the system of nonlinear equations:

$$\begin{cases} M(\varepsilon_1,\varepsilon_2) - \overline{M} = 0 \\ N(\varepsilon_1,\varepsilon_2) - \overline{N} = 0 \end{cases} \tag{1}$$

which can also be written as:

$$\begin{cases} r(\varepsilon_1,\varepsilon_2) = 0 \\ s(\varepsilon_1,\varepsilon_2) = 0 \end{cases} \tag{2}$$

Once the approximate values ε_1 and ε_2 of the (i-1)th step are known, (in the first step, for example, the values obtained from the stresses in linear elasticity can be assumed) we have, for the ith step:

$$\left| \begin{array}{c} \varepsilon_1 \\ \varepsilon_2 \end{array} \right|_i = \left| \begin{array}{c} \varepsilon_1 \\ \varepsilon_2 \end{array} \right|_{i-1} - \left| \begin{array}{cc} (\frac{\delta r}{\delta \varepsilon_1})_{i-1} & (\frac{\delta r}{\delta \varepsilon_2})_{i-1} \\ (\frac{\delta s}{\delta \varepsilon_1})_{i-1} & (\frac{\delta s}{\delta \varepsilon_2})_{i-1} \end{array} \right|^{-1} \left| \begin{array}{c} r_{i-1}(\varepsilon_1,\varepsilon_2) \\ s_{i-1}(\varepsilon_1,\varepsilon_2) \end{array} \right| \tag{3}$$

In (3) the values of the functions r and s and of the Jacobian derivatives are evaluated from the normal strains of step (i-1). The derivatives evaluated as finite difference ratios after assigning small variations of ε_1 and ε_2 and calculating the corresponding function values. This is the Newton-Raphson method, which usually gives good convergence after three or four iterations.

RCAFAG: On the Stability Analysis of Reinforced Concrete Arches and Frames   115

The new stiffnesses corresponding to each of the section elements are found as follows:

(a) *Bending stiffness.* As the elements are very short, stiffness is considered constant for each element. The stiffness of the ith step is determined as that which, with the moment of the (i-1)th step, gives the relative rotations between the two extreme sections equal to the effective one of the (i-1)th step:

$$(\int_0^1 M_{i-1}(x)dx)/(EI)_i = \int_0^1 (d\varphi/dx)_{i-1} dx \qquad (4)$$

In (4) 1 is the length of the element; $d\varphi/ds$ is the curvature obtained from the normal strains of the (i-1)th step.

(b) *Axial stiffness.* Likewise:

$$(\int_0^1 N_{i-1}(x)dx)/(EA)_i = \int_0^1 \varepsilon_{o\ i-1}(x)dx \qquad (5)$$

In (5) ε_o is the normal strain corresponding to the barycentric axis of the elements at the (i-1)th step.

FIELD OF APPLICATION

- Geometrical. The analysis concerns plane systems made up of variable sections, curved axis beams.

- Materials. Nonlinear elastic composite material reinforced concrete is considered in every analysis.

- Analysis capabilities. The programme executes first and second order analysis.

- Loadings. Selfweight, external loads (point and line loads), residual stresses (thermal loadings, distortions, pre-strains, creep, shrinkage).

PROGRAM DESCRIPTION

- Method: finite element flexibility method.

- Type of elements: beams.

- Programme structure:
    user comfort: · user facilities: data input from keyboard, or partial or total reuse of data from previous execution; interactive use; direct exploitation of results (forces and displacements) or storage in files for later use.

    · automatic mesh generation and node and element numbering.

    · postprocessors: drawing of the results, stress calculation, reaction calculation.

    R & D: · language of the program: BASIC.

    · new developments: capacities for introduction of new subroutines to describe different constitutive laws and different convergence criteria.

HARDWARE COMPATIBILITIES

- Minimal configuration of materials required: 24 K RAM computer, printer and plotter.

- Type of computer and peripherals: Hewlett-Packard.

- Media: available in magnetic tape.

EXAMPLES OF APPLICATION

Reinforced Concrete Portal Frame. Nonlinear Analysis

The example in Fig. 1 shows the nonlinear limit analysis of an r.c. portal frame.

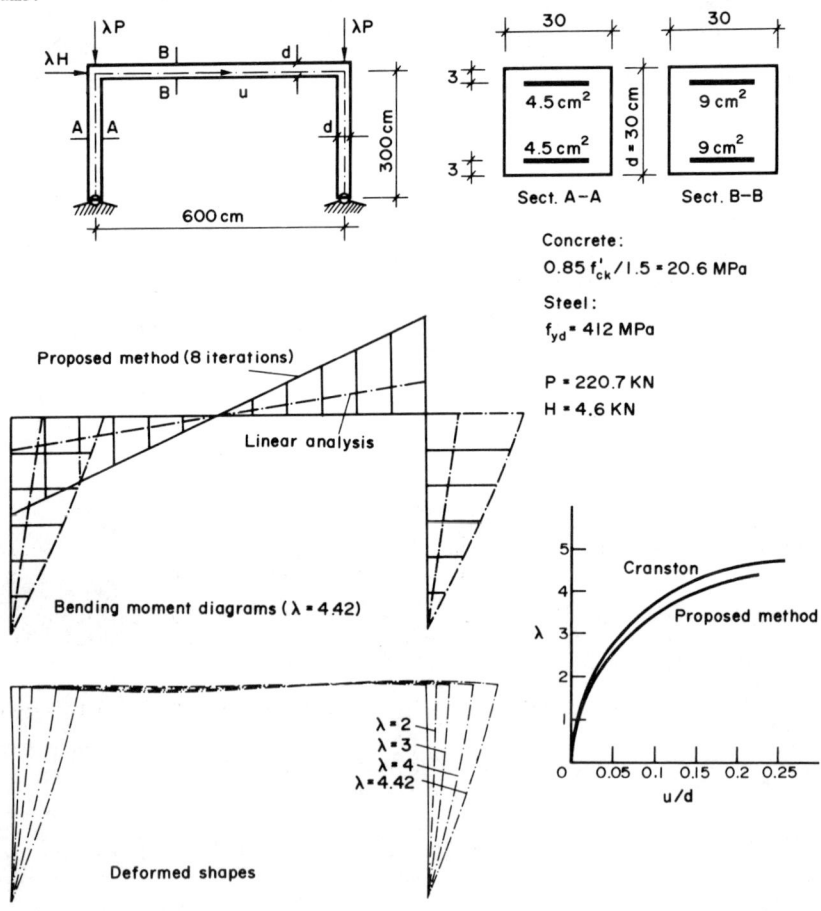

Fig. 1. Nonlinear analysis of Cranston's portal frame.

The analysis, performed with RCAFAG, was used as a test for the program itself.

It takes into account both geometrical nonlinearities and the nonlinearities of reinforced concrete material. The constitutive laws proposed in the FIP-CEB Model Code were used for the materials steel and concrete. This analysis is hence held to be particularly significant.

Each column was subdivided into 6 elements and the upper cross beam into 12 elements, giving 24 elements in all. The degree of indetermination was 1.

The figure shows the geometrical characteristics of the structure, and of the materials and the loads. The bending moments, deformed shapes at various proportionally increasing load levels and the normalized horizontal displacement of the upper cross beam as a function of the load multiplier are shown.

The analysis was performed with RCAFAG on HP 9831A computer, HP 9871A printer, HP 9872A plotter.

Reinforced Concrete Bridge Arch. Nonlinear Analysis

An illustration of the analysis used for the determination of the critical load of a fixed bridge arch in r.c., submitted to self-weight and to two uniformly distributed loads, representing codified bridge loads.

The arch, which is analyzed with RCAFAG, is of variable height with the law $h = h_c/\cos\psi$, where h_c is the height at the apex section and ψ is the characteristic angle of the generic section; the width is constant at 5.00 m. It is made of concrete ($E = 2.10^7$ kN/m^2; material unit weight = 25 kN/m^3; Poisson ratio = 0.2). The constitutional laws proposed in the FIP-CEB Model Code are used for the steel and the concrete.

Three types of analysis were performed: linear elastic, only geometrically nonlinear, nonlinear geometrically and for the materials: the comparative results are shown in Fig. 2, with, for a section at left-hand base, the routes of vectors M and N in the respective analyses within the field of safety up to the point of encounter with the interaction diagram.

The considerable differences in the maximum load multiplier λ, and hence in the degree of safety, may be seen.

The problem relates to safety analysis of deformable r.c. arches and is applicable in the field of civil engineering, in particular to the construction of large bridges and the analysis of already existing arch bridges.

In the example shown here the structure is subdivided into 60 beam elements liable to deformation by normal and shear forces. There are 59 joints. The degree of indetermination is 3.

The equipment used in the calculation is HP 9831A computer, HP 9871A printer, HP 9872A plotter.

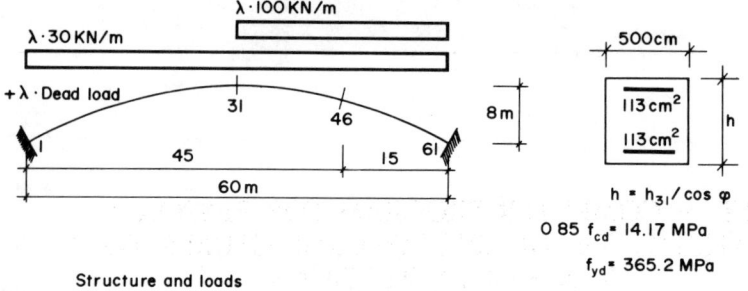
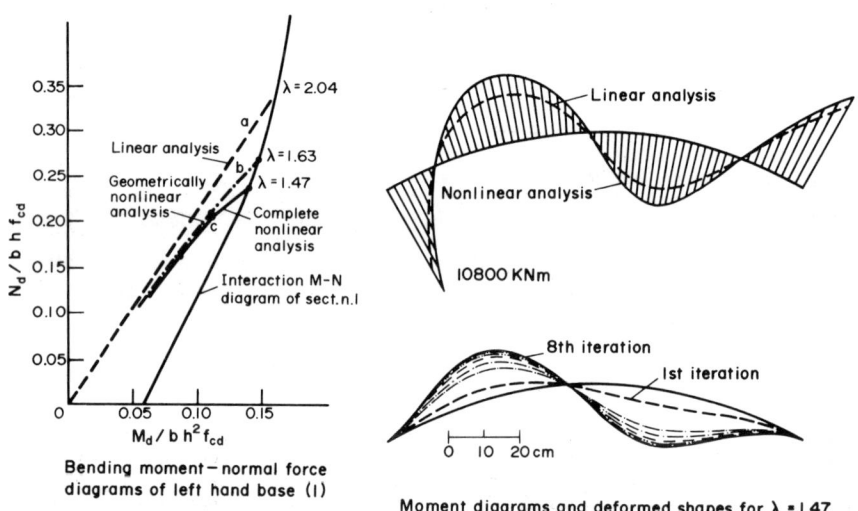

Fig. 2. Analysis of r.c. bridge arch

REST: A COMPUTER PROGRAM FOR RESPONSE STATISTICS OF DISCRETIZED STRUCTURES TO RANDOM EXCITATIONS

C. W. S. To

Department of Mechanical Engineering, The University of Calgary, Calgary, Alberta, Canada T2N 1N4

ABSTRACT

REST is a finite element program for the computation of REsponse STatistics of linear structures to non-stationary random excitations. Various finite elements included are tapered beams, two-dimensional solid, three-dimensional solid, plate bending and thin shell. User-defined elements can be incorporated readily when required. Every non-stationary random excitation applied is represented as a product of a deterministic modulating function and a Gaussian process which is not necessarily white. When the latter is white the solution in the time domain is exact.

INTRODUCTION

REST is a system of computer programs for determining the natural frequencies and mode shapes, response statistics, and various hazard functions and first passage probabilities for type-D crossings of general linear structures. The programs are written in Fortran 77. Their development to date has been on a CDC cyber 175 machine.

The analysis technique used is of two parts. The first part is concerned with the spatial domain discretization while the second part deals with the time domain solution. In the spatial domain discretization the finite element displacement method is used. In the time domain solution various response statistics that are required for the computation of hazard functions and first passage probabilities for type-D crossings are obtained in closed form. In the finite element displacement method the structure is represented by an assemblage of discrete elements which are connected to one another at the nodal points. A collection of typical elements are provided. In addition, more finite elements can be included by the users without program modification. Essentially, this part of the system of programs is based on SPADAS 1 (see reference 1). This implies that every feature that SPADAS 1 possesses is preserved in REST.

In practice the system of computer programs resides on disc or magnetic tape in compiled form. The first stage of the analysis is prepossessing the input data. The second stage is solving the eigenvalue problem of the complete structure. The third stage is performed interactively such that analysis control instructions can be selected appropriately and conveniently. The final stage is concerned with the management of output data.

THEORETICAL BACKGROUND

The linear time-invariant structure is discretized by the finite element displacement method and can be described by the matrix equation of motion

$$M\ddot{X} + C\dot{X} + KX = P \qquad (1)$$

where M, C and K are the assembled mass, damping and stiffness matrices of the system respectively while $\ddot{X}$, $\dot{X}$, X and P are, respectively, the random acceleration, velocity, displacement and applied force vectors.

The uncoupled form of equation (1) is

$$\ddot{Q}_r + (\lambda_m + \lambda_k \omega_r^2)\dot{Q}_r + \omega_r^2 Q_r = F_r(t), \qquad r = 1,2,3,\ldots,n \qquad (2)$$

The non-stationary random excitation considered here can be expressed as

$$F_r(t) = \psi_r(t)\eta_r(t)$$

where $\psi_r(t)$ and $\eta_r(t)$ are the deterministic modulating function and the stationary process. The components of the response vector are

$$X_j(t) = \sum_{r=1}^{n} (_r r_j) Q_r(t), \qquad (3)$$

where $_r r_j$ is the element of the eigenvector associated with the rth mode of the discretized structure.

Based on the time domain approach and with reference to equation (2)

$$Q_r(t) = \int_0^t h_r(t-\tau) F_r(\tau) d\tau, \text{ or, in detail,}$$

$$Q_r(t) = \frac{1}{\Omega_r} \int_0^t e^{-\zeta_r \omega_r (t-\tau)} \sin \Omega_r (t-\tau) \psi_r(\tau) \eta_r(\tau) d\tau, \qquad (4)$$

where $\Omega_r = \omega_r (1-\zeta_r^2)^{\frac{1}{2}}$, $2\zeta_r \omega_r = \lambda_m + \lambda_k \omega_r^2$, and λ_m and λ_k are defined by the following relation

$$C = \lambda_m M + \lambda_k K .$$

Response Statistics

By making use of the theory in Section 2 and results in Sections 6 and 7 of reference 2, the covariance of displacement responses $X_j(t)$ and $X_k(t)$ is

$$E < X_j(t) X_k(t) > = \int_{-\infty}^{\infty} S_{jk}(t,\omega) d\omega \qquad (5)$$

where $S_{jk}(t,\omega)$ is the evolutionary cross-spectral density function of the nodal displacements $X_j(t)$ and $X_k(t)$ at time **t**.

The detailed expression of equation (5) is given in reference 3.

Similarly, the closed form time-dependent covariance of displacement $X_j(t)$ and velocity $Y_k(t) = \partial X_k(t)/\partial t$ can be written as

$$E < X_j(t) \, Y_k(t) > \; = j \int_{-\infty}^{\infty} \omega \, S_{jk}(t,\omega) \, d\omega \tag{6}$$

and the closed form covariance of velocity responses $Y_j(t) = \partial X_j(t)/\partial t$ and $Y_k(t) = \partial X_k(t)/\partial t$ is given as

$$E < Y_j(t) \, Y_k(t) > \; = \int_{-\infty}^{\infty} \omega^2 S_{jk}(t,\omega) \, d\omega \; . \tag{7}$$

The detailed expressions of equations (6) and (7) are included in reference 3.

First Passage Probability for Type-D Crossings

The first passage probability for type-D barrier based on the Poisson process assumption is given as reference 4.

$$L_D(t) = \exp\left[-\int_0^t \alpha_D(u) \, du\right], \tag{8}$$

where $\alpha_D(t) = 2\nu_b(t)$ and

$$\nu_b(t) = \frac{1}{2\pi} \left[\lambda_2(t)/\lambda_0(t)\right]^{\frac{1}{2}} \exp\left[-\frac{b^2}{2\lambda_0(t)}\right]$$

in which

$$\lambda_i(t) = \int_{-\infty}^{\infty} \omega^i S_{jj}(t,\omega) \, d\omega \; , \quad i = 0,2$$

and

$$\lambda_1(t) = j \int_{-\infty}^{\infty} \omega \, S_{jj}(t,\omega) \, d\omega \; ,$$

and b is the barrier level.

Various considerations for type-D crossings can also be found in reference 4. For brevity they are not included here.

PROGRAM STRUCTURE

The original operational guidelines that governed the structure of the system of computer programs were that the system was to:

(a) be aimed at random analysis of response of general linear structures excited by a variety of transient disturbances such as earthquakes, blast waves of explosion and atmospheric turbulence,

(b) span a considerable range of linear structures such as aerospace structures and naval structures,

(c) have an adequate, but not necessarily complete, library of finite elements, and,

(d) be open ended for updating.

With the above range the system is divided into two parts, namely, the spatial discretization and the time domain solution. The technique used in the first part is the finite element displacement method. In this method the structure is represented by an assemblage of discrete elements which are connected to one another at the nodal points. A series of finite elements is provided and shown in Fig. 1.

Additional elements can be incorporated by the user without modifying the system. The second part of REST involves with the time domain solutions for response statistics and first passage probabilities. The response statistics, briefly mentioned in Section 2 above, are the variances and covariances of nodal displacements, variances and covariances of nodal velocities, covariances of nodal displacements and nodal velocities. The first passage probabilities considered in the system are concerned with type-D crossings. In this type of barrier the bounds comprise of a pair of lines, $X_j(t) = b$ and $X_j(t) = -b$, and the safe region is defined by $|X_j(t)| < b$.

The aforementioned two parts are further divided into four stages of analysis. The first stage is concerned with pre-processing the input data which have been organized as an input file. The second stage deals with the eigenvalue solution of the complete structure. The third stage consists of a series of analysis control instructions that are input interactively so that the computed results can be checked. The final stage is concerned with output data management. With the IMSL (International Mathematical and Statistical Library) REST can provide graphs interactively.

The language used in the system is standard Fortran 77. Three directions of further developments are in progress. One is concerned with analysis technique in which the frontal method (see reference 5) is to be incorporated. The second direction deals with increasing the element library capability. The third direction is to do with considering other types of crossings for the first passage probability computation.

HARDWARE COMPATIBILITIES AND FIELD OF APPLICATION

Currently, the mainframe computer used is CDC cyber 175 with NOS 2.1 at the University of Calgary. The system of programs resides on disc or magnetic tape in compiled form.

Typical elements included in REST are tapered beams, two-dimensional solid, three-dimensional solid, triangular, rectangular plate bending, and thin shells. These elements have already been illustrated in Fig. 1. For brevity they shall not be repeated here. The material considered is linear elastic and isotropic. The eigenvalue problem of any linear structure can be solved. Response statistics and first passage probability of any linear structure that is excited by various transient disturbances modelled as non-stationary random excitation can be computed. The type of excitations is exemplified in Fig. 2.

EXAMPLE

The example considered here is a physical model of a class of mast antenna structures. The structure is assumed to be rigidly clamped at the base. The discretized physical model is shown diagrammatically in Fig. 3. This is the case considered in reference 2. The antenna was approximated by two discrete masses including rotary inertias lumped at node number 14 and 16. The beam and discrete mass finite elements used are TB5 and DM3 presented in reference 6. The input data for the structure can be found in reference 2. The envelope modulated random excitation is applied at the base of the structure, that is, node number 1 of Fig. 3(b). Thus, the governing equation of motion in matrix form is

A Computer Program for Response Statistics   123

$$\begin{bmatrix} M_{yy} & M_{yx} \\ M_{xy} & M_{xx} \end{bmatrix} \begin{Bmatrix} \ddot{Y} \\ \ddot{X} \end{Bmatrix} + \begin{bmatrix} C_{yy} & C_{yx} \\ C_{xy} & C_{xx} \end{bmatrix} \begin{Bmatrix} \dot{Y} \\ \dot{X} \end{Bmatrix} + \begin{bmatrix} K_{yy} & K_{yx} \\ K_{xy} & K_{xy} \end{bmatrix} \begin{Bmatrix} Y \\ X \end{Bmatrix} = \begin{Bmatrix} F_y \\ 0 \end{Bmatrix} \quad (9)$$

where (M_{xx}), (C_{xx}) and (K_{xx}) are the mass, damping and stiffness matrices of the constrained structure; $\{X\}$ and $\{Y\}$ are the random displacement response and prescribed random displacement column matrices, respectively, and $\{F_y\}$ is the column matrix of unknown forces causing the displacement $\{Y\}$.

Consider the second of equations (9). Then

$$(M_{xx})\{\ddot{X}\} + (C_{xx})\{\dot{X}\} + (K_{xx})\{X\} = \{F_x\} \quad (10)$$

where $\{F_x\} = -(M_{xy})\{\ddot{Y}\} - (C_{xy})\{\dot{Y}\} - (K_{xy})\{Y\}$. Referring to equation (2), one has, for envelope modulated random excitation

$$F_r(t) = \frac{1}{m_{rr}} \lfloor {}_r r_1 \; {}_r r_2 \; \cdots \; {}_r r_n \rfloor \{F_x\}, \quad (11)$$

where m_{rr} and ${}_r r_j$ are, respectively, the diagonal element of the diagonalized mass matrix and element of the eigenvector associated with the rth mode of the constrained structure in equation (10). As the terms associated with the applied velocity and acceleration are small compared with that associated with the prescribed displacement because the elements of (K_{xy}) are usually several orders higher than those of (M_{xy}) and (C_{xy}), the excitation vector reduces to

$$\{F_x\} = -\lfloor K_{xy} \rfloor \{Y\}, \quad (12)$$

where in this example $\{Y\} = [\tilde{y}(t)0]^T$, in which the superscript T denotes "the transpose of". Operating on equation (12) leads to

$$\{F_x\} = -\lfloor k_{31} \; k_{41} \; k_{51} \; k_{61} \; 0 \; 0 \; \cdots \; 0 \rfloor^T \tilde{y}(T) \quad (13)$$

$n \times 1 \qquad\qquad 1 \times n$

where $k_{31}, \ldots$, are the elements of the assembled stiffness matrix. In this example $n = 48$.

Applying equation (11) gives

$$F_r(t) = -(1/m_{rr}) \lfloor ({}_r r_1) k_{31} + ({}_r r_2) k_{41} + ({}_r r_3) k_{51} + ({}_r r_4) k_{61} \rfloor \tilde{y}(t)$$

$$= \psi_r(t) \eta_r(t),$$

where $\psi_r(t)$ is the envelope modulating function.

$$\eta_r(t) = -(1/m_{rr}) \lceil ({}_r r_1) k_{31} + \ldots + ({}_r r_4) k_{61} \rfloor \gamma(t): \quad (14)$$

that is, $\tilde{y}(t) = \psi_r(t)\gamma(t)$. $\gamma(t)$ is the stationary part of the random displacement applied at the base of the structure. The spectral density of $\gamma(t)$ is $S_{\gamma\gamma}(\omega)$: that is

$$E\langle \eta_r(\tau_1) \eta_s(\tau_2) \rangle = \int_{-\infty}^{\infty} \frac{1}{m_{rr}} \lfloor ({}_r r_1) k_{31} + \ldots + ({}_r r_4) k_{61} \rfloor$$

$$\times \frac{1}{m_{ss}} \lfloor ({}_s r_1) k_{31} + \ldots + ({}_s r_4) k_{61} \rfloor e^{j\omega(\tau_1 - \tau_2)} S_{\gamma\gamma}(\omega) d\omega.$$

The modulating function $\psi_r(t)$ is plotted in Fig. 4. Figure 5 presents results of probabilities of no D-crossing for various barrier levels at $j = k = 29$. This is associated with the displacement response statistics at node 11.

REFERENCES

1. Petyt, M. and A. Y. Abdel-Rahman (1974). Institute of Sound and Vibration Research, University of Southampton. SPADAS 1 Theoretical Manual.

2. To, C. W. S. (1984). *Journal of Sound and Vibration*, 93(1), 135-156. Time-dependent variance and covariance of responses of structures to non-stationary random excitations.

3. To, C. W. S. (1983). Rept. No. 275, Department of Mechanical Engineering, The University of Calgary. Response statistics of structures to non-stationary random excitation.

4. To, C. W. S. (1984). Rept. No. 296, Department of Mechanical Engineering, The University of Calgary. First-passage problem of ship mast antenna structures to non-stationary random excitation.

5. Irons, B. and S. Ahmad (1980). *Techniques of Finite Elements*. Chichester: Ellis Horwood Ltd.

6. To, C. W. S. (1979). *Journal of Sound and Vibration*, 63, 33-50. Higher order tapered beam finite elements for vibration analysis.

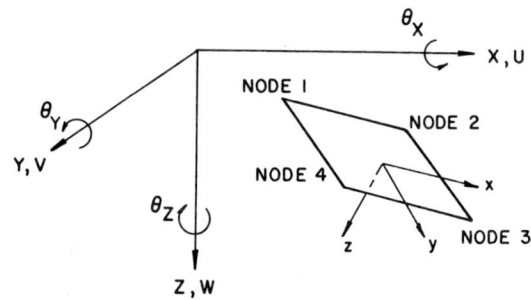

(a) Isotropic, rectangular flat shell element.

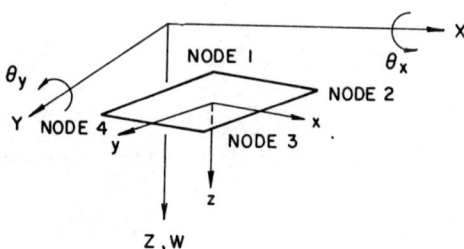

(b) Two-dimensional isotropic rectangular plate bending element.

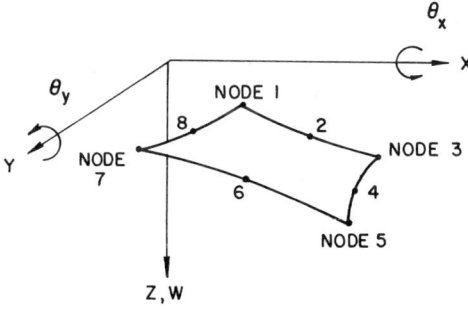

(c)   Two-dimensional isoparametric plate bending element.

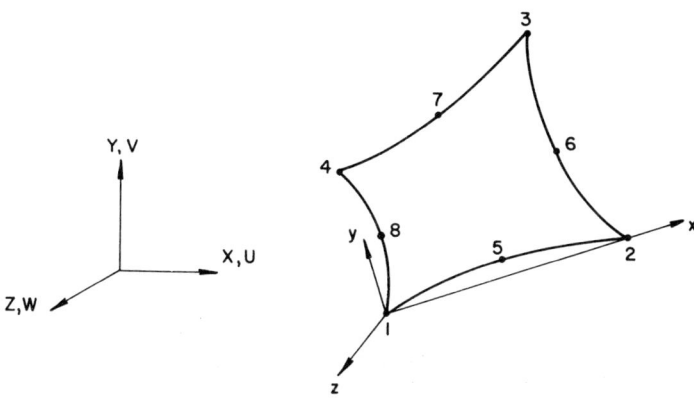

(d)   Isoparametric thick plate element, with bending and
        membrane displacements.

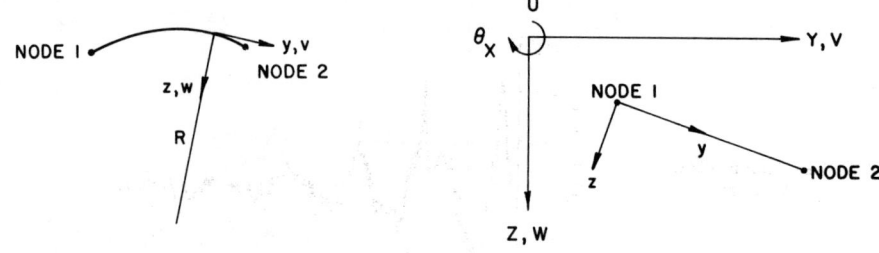

(e) Finite strip singly curved shell element.

(f) Finite strip, flat shell element.

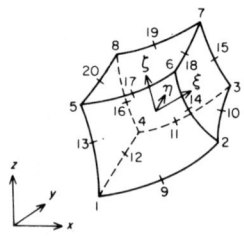

(g) 20-node isoparametric solid element.

Figure 1. Some Typical Elements in REST.

A Computer Program for Response Statistics

Ship Shock Time History

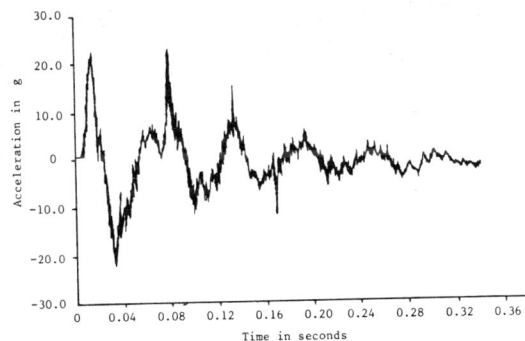

Accelerogram from El Centro
Earthquake, May 18, 1940
(North-South component)

Figure 2. Typical Random Excitations.

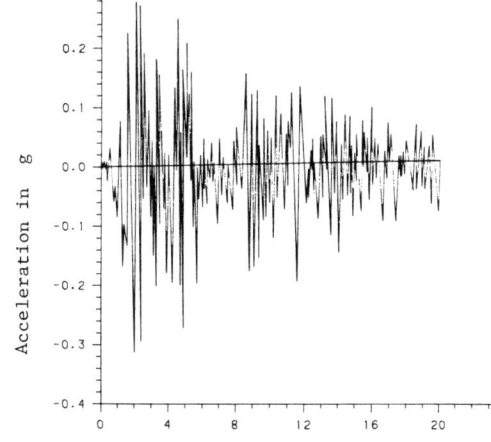

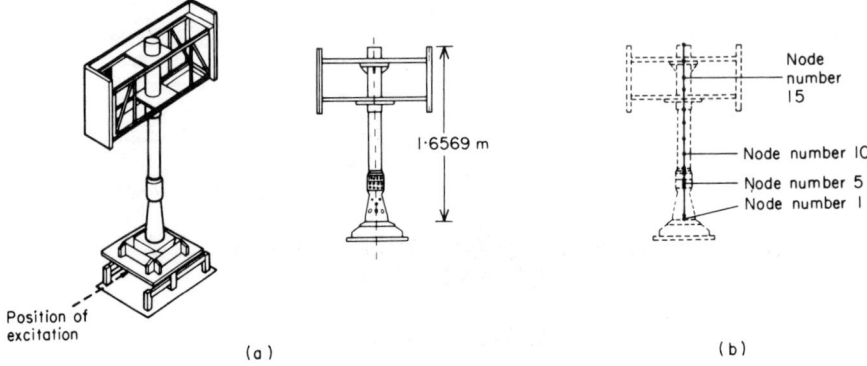

Figure 3. The Finite Element Representation of The Physical Model of a Mast Antenna Structure. (a) The physical model; (b) Finite element idealization of (a).

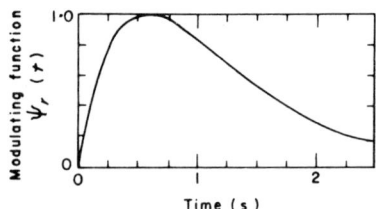

Figure 4. The Modulating Function of The Excitation.

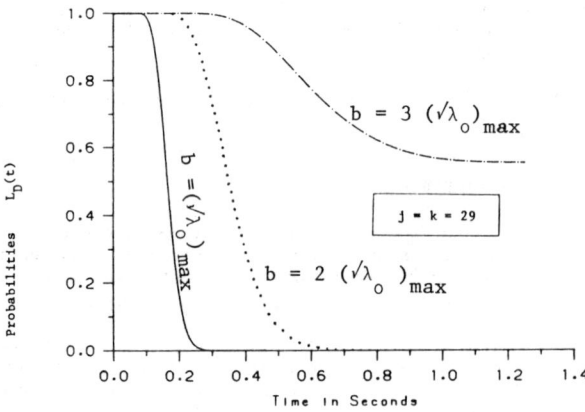

Figure 5. Probabilities of No D-Crossing for Various Barrier Levels.

SIMP — DESIGN SENSITIVITY, ADVANCED CAE AND CAT

K. Zimmermann

AEO Ltd. (Aerospace Engineering Office Ltd.), Universitätstr. 87, POB 200, CH-8033 Zurich, Switzerland

ABSTRACT

SIMP is a well-established, advanced CAE (Computer Aided Engineering) and CAT (Computer Aided Testing) package which includes design sensitivity for structural systems. It has been used for many years in space projects. SIMP runs in interactive or batch processing mode. SIMP is developed at AEO Ltd., Zurich, Switzerland, and marketed worldwide by FIDES Trust Company, Zurich, Switzerland.

SIMP enables a user friendly and efficient design sensitivity analysis for design investigation and optimization, as well as to adapt FE models to test results. By using the information of design sensitivity, the engineer gets immediate interactive answers to any design changes and is thus enabled to yield optimum design guidelines. Sensitivity analysis opens a new, more efficient and cost effective area in the field of structural design analysis.

The program also provides the well-known features of pre- and postprocessors, doing powerful mesh generation, extensive editing, displaying and/or plotting, numerical and graphical data evaluation. Furthermore SIMP includes many other general purpose CAE modules.

SIMP's CAT capabilities include automatic selection of relevant degrees of freedom for dynamic tests.

THEORETICAL BACKGROUND

SIMP has been written specifically to take advantage of the latest developments in computer technology and its developers have succeeded in producing a fast and efficient system which is not only comprehensive but also has a simple and straightforward user interface.

Increasing demands in finite element analysis led to the development of various pre- and postprocessors. They reduced the effort spent for preparing the FE model considerably. These pre- and postprocessors - which sail under the letters CAE - are well separated from the FE code (Fig. 1).

Refined and more complex FE models increase the turn around time, as well as the time for the evaluation of data. From a financial point of view, it also reduces the possibility of keeping the model sufficiently updated. These processors

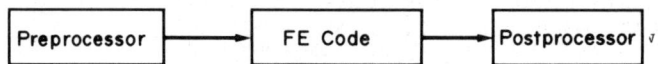

usually offer adequate graphical presentation of the FE results. However, there is still a lot of time spent by the engineer evaluating numerically these data, documenting the results and conducting e.g. a detail stress analysis, correlating them to test results or even fitting the FE model to measured values. As SIMP offers all these analytical tools - let them be summarized as advanced CAE - it becomes a very powerful software package for all structural engineers. Although an advanced CAE package requires some more interfaces than conventional pre- and postprocessors, they can still be operated interactively (Fig. 2).

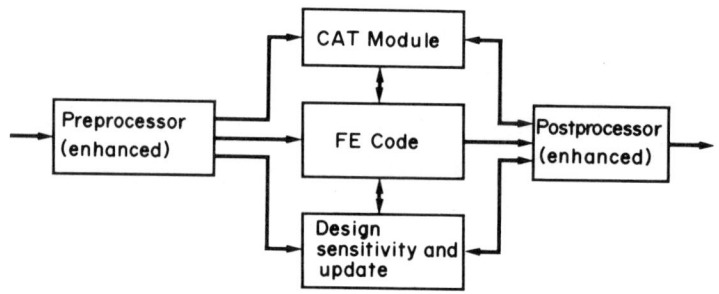

FIELD OF APPLICATION

Method

Sensitivity algorithms implemented in SIMP are based on differential design investigations of displacements, stresses, eigenvalues and eigenvectors due to a single design change of a FE model parameter.

SIMP's design sensitivity algorithms are - contrary to most other sensitivity algorithms - physically interpretable and easy to apply since the topological description of the mathematical idealization i.e. single parameters (physical properties such as mass density, stiffness of an element or subassembly, area moments of inertia, cross-sectional area, shear stiffness, etc.) are analysed.

Modules

A sensitivity investigation may be performed using a rigid format or calling user selected modules interactively. The following modules are implemented:

- Influence of a user defined design change on FE results (design investigation)
- Cross-orthogonality and automatic reordering of modes (correlation of mode shapes)
- Model comparison and comparison of test and analysis
- Adaptation of analytical models to objective functions or test results (model

fitting)
- Prospective update of FE results (displacements, stresses, eigenfrequencies, mode shapes)

Analysis Capabilities

SIMP's design sensitivity algorithms support static and dynamic systems especially:

- Static displacements
- Strains
- Stresses
- Eigenvalues
- Generalized masses
- Eigenvectors
- Cross-orthogonality

Element Types

Differential sensitivity algorithms may be implemented for all structural elements of a FE program. Actually all beam elements and triangular or quadrilateral elements with independent membrane, bending and transverse shear properties may be used for design investigations.

PROGRAM DESCRIPTION

Pre- and Postprocessing

Mesh generation. Using SIMP's mesh generation module the user only needs to define a few key words to mesh surfaces and solids with any mesh density and spacing. In addition, there are facilities for generating nodes on straight lines and circular areas.

The mesh generation module allows substructuring, in order to generate subsystems and merge them to a main structure.

Substructures may be duplicated, translated, rotated and merged to the main structure.

Graphics. For the most graphic modules, SIMP offers a display on screen and/or a plot.

- Structural plot.

  SIMP contains a number of graphical facilities to aid the user in viewing and checking his model: Selection of elements to be viewed, deformed shape, shrinked elements, zooming, plan-, isometric- or central projection, particular or general labeling of grids and/or elements, element or grid modifications with cross-hair cursor.

- Contour plot.

  Stresses, temperatures, displacements and modes can be shown in contour plot. All projection types of the structural plot are enabled. Contour lines can be requested on the undeformed and the deformed shapes.

- Sequence plot.

    Stresses, temperatures, displacements and modes etc. can be plotted against grid or element sequences for user selectable load cases.

- Vector plot.

    User selectable grid and degree of freedom (dof) sequence can be involved in displaying deformed shapes and modes, even those dof's measured. Correlation of measured and analysed dof's is possible.

Figure 3 illustrates some display options.

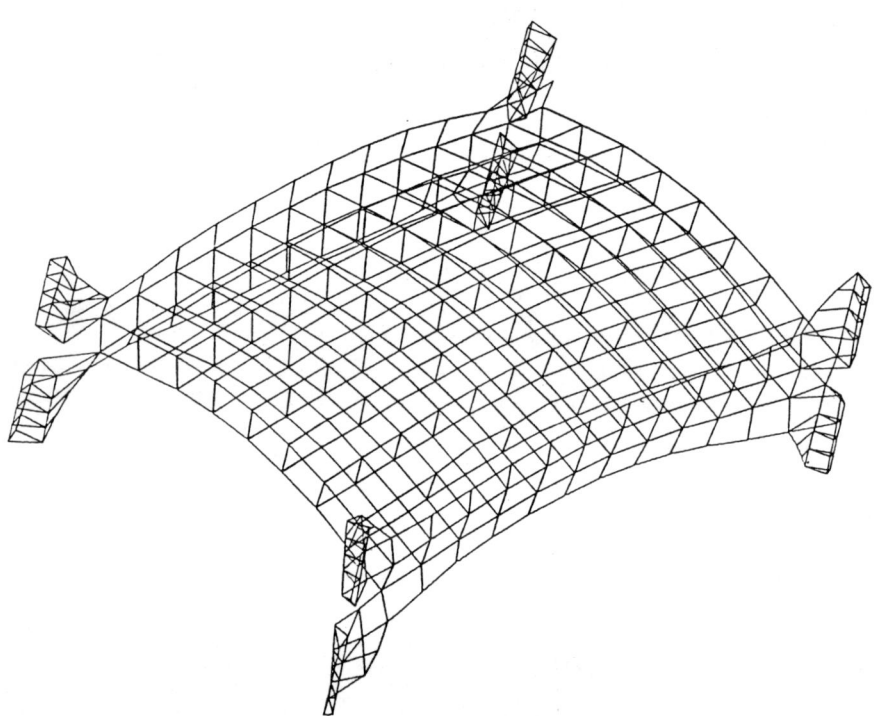

Fig. 3. Some graphical display options offered by SIMP

READY-TO-PRINT OUTPUT

Tailoring of numerical output by the user such as for

- Displacements (selectable grid set, dof's, load cases, search for maximum displacements, labeling)
- Stresses (selectable element set, stress type, such as equivalent stress, etc., load cases, search for maximum stresses or stress range, labeling)
- Dynamic model comparison, see Table 1

SIMP: Design Sensitivity, Advanced CAE and CAT

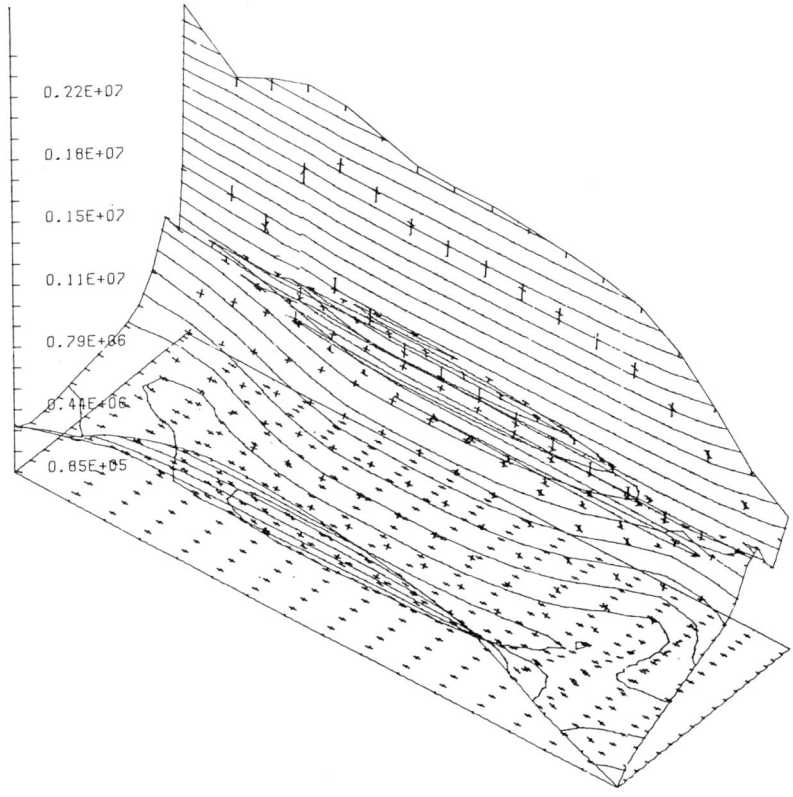

Fig. 3. Some graphical display options offered by SIMP

Advanced CAE

Sensitivity analysis. Differential sensitivity of structural systems is used for investigation of design changes, accounting for design updates using the results of the original system and adaptation of mathematical models to test results. Various approaches to find optimum identification of parameters, to adapt a mathematical model are implemented.

Having these design analysis tools at hand, the engineer can produce timely updates of analysis results, provide immediate answers to any design changes in consideration, saving computer cost. He gets much better analytical information and is thus enabled to yield optimum design guidelines.

Modules:

- Design investigation
- Cross-orthogonality
- Correlation and automatic reordering of modes
- Model comparison
- Design optimization

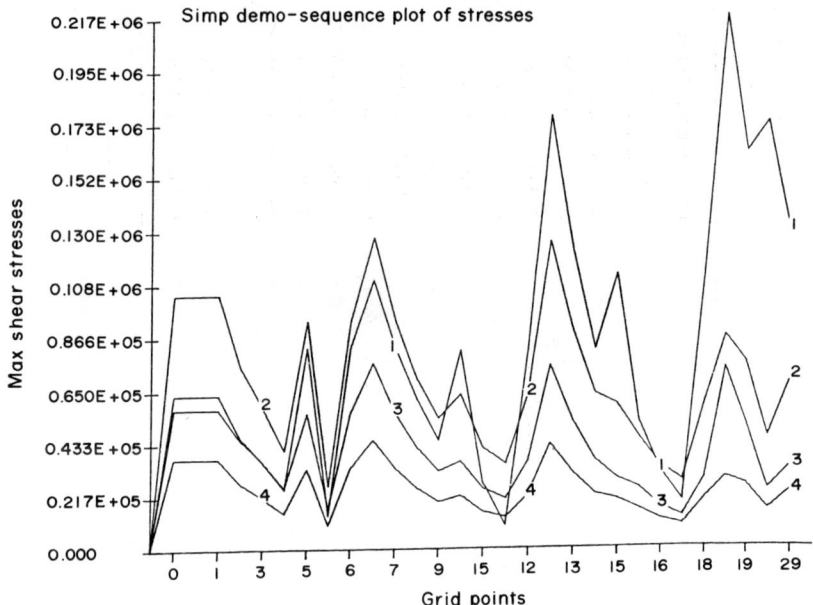

Fig. 3. Some graphical display options offered by SIMP

- Model fitting
- Prospective updating

Automatic selection procedure for dynamic degrees of freedom. SIMP has the facility to automatically select relevant degrees of freedom for dynamic tests. The selection is done neglecting rotational degrees of freedom, redundant rigidly coupled degrees of freedom and degrees of freedom which are insignificant for describing the derived mode shapes.

The efficiency of the approach was demonstrated using a mathematical model of the spacecraft dummy structure (FLECS). As has been demonstrated, even experienced test engineers reach only about 60% or less of the significant and non-redundant degrees of freedom.

A few minutes of computer time yielding an optimized set of measurement points can be traded against hours or days with discussions and uncertainties defining measurement points by experience.

Vector and matrix operations. Normalization of vectors, superpositioning, linear combinations, computations of generalized mass and stiffness, etc.

Section properties. Any arbitrary cross section may easily be generated by SIMP's section property module. After generation, cross-sectional area, properties and main axes will be computed and displayed or plotted. A NASTRAN PBAR card may be written into the input data deck.

Best fit paraboloid computing. SIMP enables the computation of best fit paraboloids of thermally and mechanically distorted parabolic antennas. The antenna, defined in rectangular coordinates may have up to two symmetry planes.

TABLE 1 Ready-to-print table provided by SIMP

TABLE 1		COMPARISON PHYSICAL MODEL -- MATH.MODEL- 1										
MODE NO.		FREQUENCIES			GENERALIZED MASSES			KINETIC ENERGIES			CROSS ORTHOGONALITY	
T	A	T	A	RATIO	T	A	RATIO	T	A	RATIO	D	M
3	1	27.797	27.110	0.975	45.880	37.310	0.813	6.99BE+05	5.413E+05	0.774	0.7286	0.0000
2	2	26.921	27.141	1.008	34.832	34.498	0.990	1.983E+05	5.016E+05	1.007	0.9997	0.0001
1	3	26.044	31.720	1.281	52.031	73.057	1.400	6.966E+05	1.447E+06	2.007	0.5935	0.6835
4	4	31.571	35.319	1.002	37.843	65.073	1.720	8.927E.05	1.6035E+06	1.795	0.8363	0.5461
5	5	39.278	39.640	1.009	18.870	20.134	1.067	5.747E+05	6.245E+05	1.087	0.9967	0.0031
6	6	49.681	47.570	0.958	19.037	19.970	1.049	9.275E+05	8.920E+05	0.962	0.9949	0.0801
7	7	50.255	50.454	1.004	84.813	54.520	0.643	4.220E+06	2.740E+06	0.648	0.9637	0.0498
8	8	66.913	65.340	0.976	83.791	55.971	0.668	7.406E+06	4.717E+06	0.637	0.9656	0.2583
9	9	99.092	98.089	0.990	109.493	107.139	0.970	2.122E+07	2.035E+07	0.959	0.9987	0.0347
10	10	121.163	122.234	1.009	62.248	63.614	1.022	1.804E+07	1.876E+07	1.040	0.9989	0.0246
SUM AND RATIOS OF SUM					548.841	531.086	0.968	5.518E+07	5.217E+07	0.945		
MEAN VALUES				1.017			1.035			1.098	0.9076	0.1680
STANDARD DEVIATIONS				0.073			0.324			0.473	0.1426	0.2496

In order to increase numerical accuracy, the third coordinate is recomputed after antenna shape parameters have been defined.

3D stresses (temperatures and boundaries). Computes 3D equivalent stresses of a cross-section modelled by 2D elements as well as based on temperature distribution and boundary conditions.

Large displacement motion analysis. Large displacement motion analysis superimpose rigid body motions and elastic deformations. In this method, the amplitudes of the modes, their eigenfrequencies and shapes and the rigid body velocities are used to compute large displacement motions of an elastic structure. If the shape of the model (e.g. launcher or spacecraft) is given, the relative movement and clearance between the two separating bodies is the direct and important result.

Other features. A number of other capabilities are implemented or inherent in the design of SIMP. The content of SIMP is under continuous review and extensions and improvements are made as developments are proved and verified.

Interface

SIMP is specially interfaced to NASTRAN (MSC, Cosmic or MARC) but can also be used for other FE codes by converting the model data since SIMP employs its own database.

Processing Mode

SIMP is designed for interactive use. It employs free format input, optionally with keywords. However, single steps or a complete command sequence may be processed as rigid format. In addition, SIMP runs also as batch job.

Input

SIMP is menu oriented, i.e. the user may select a module and will be guided through the menu with interactive questions. Its modules are easy to learn and remember, making the system particularly attractive to the beginner or occasional user. The experienced user may work with SIMP using a command line input.

Output

Results of the sensitivity analysis may be presented by the general purpose postprocessor, especially by (numerical) ready-to-print tables or graphical display options (structure, contour- and sequence-plot). In the case of model fitting a complete FE input data deck will be established.

User Extensions

The complete program is written in FORTRAN 77. Some menus are reserved for the user in order to give the possibility to implement specific requirements. The user has easy access to SIMP's database.

HARDWARE COMPATIBILITIES

SIMP is supplied for CDC Cyber computers (NOS-operating system), DEC/VAX computers (VMS-operating system) as well as PRIME computers (PRIMOS). The program can be adapted to other computers having virtual memory capabilities.

As peripherals, graphic terminals which are compatible with Tektronix 4010 or Tektronix 4014 are required. Plotter output may be generated on any Calcomp or Graphtec series compatible device.

DOCUMENTATION

A complete description of the program is given in the user manual, which consists of:

- Program Overview
- Menu Reference Guide
- Menu Description
- Applications
- Primer
- Error List

AVAILABILITY AND LICENCES

SIMP is installed at FIDES Computer Service, Zurich, Switzerland and may be used by all FIDES customers. FIDES is operating a worldwide communications network which also allows direct access via TELEPAC or DATEX-P.

Installation and utilization are possible by acquiring a licence.

APPLICATIONS

Design Optimization

Using differential sensitivity algorithms of SIMP, finite element models can be optimized by introducing an objective function.

AEO designed the necessary reinforcements of a tundish trolley, in order to stay within stress and deformation limits. Design sensitivity analysis yielded an optimized design, keeping the reinforcements to an absolute minimum while simplifying the structural system (Figs. 4-6).

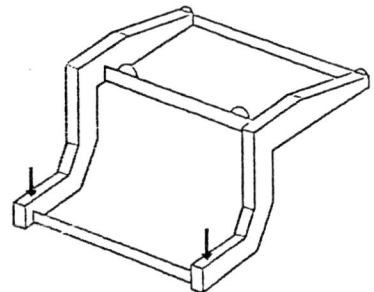

Fig. 4. Tundish trolley (overall structure).

Dynamic Model Update

The comparison of dynamic characteristics of structural systems, identified by analysis and test, usually shows considerable discrepancies (Table 2).

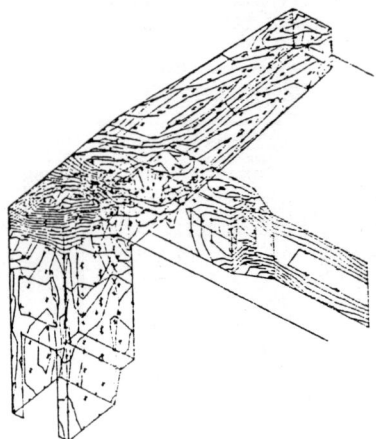

Fig. 5. Contour plot of initial stresses of tundish trolley substructure.

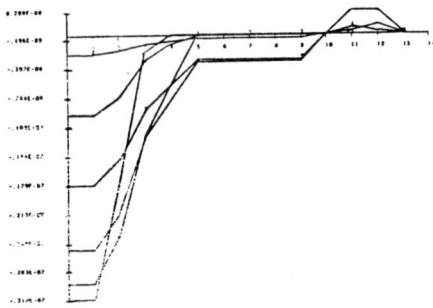

Fig. 6. Sensitivity vectors of displacements.

Updating of the FE model to test results saves time and cost of testing during design verification. In the frame of ESA contracts, AEO performed an adaptation of all modal parameters of the spacecraft dummy structure FLECS. This structural model was designed such that frequencies and location of modes coincide well with a typical satellite structure (Fig. 7).

Using SIMP's differential sensitivity algorithms the adaptation yielded overwhelming improvements in most modal parameters of the structural system.

Geometric Nonlinear Model Update (Figs. 8,9)

Under contracts of the European Space Agency ESA, AEO established the theoretical basis for the design verification of large flexible solar arrays. Approach and methods enabled to correlate the behaviour of these solar arrays under gravity forces with those of the on-orbit condition, as well as to update the mathematical model to test results. The prediction of the behaviour of these arrays is important because it has a considerable influence on the attitude and orbit control system of the spacecraft. Such flexible structures are mostly not capable

TABLE 2 Comparison of Modal Parameters. (1) Original Analytical
Model - Test Results, (2) Updated Analytical
Model - Test Results

Criteria	Orig. Math. Model	Component Mode	
		Values	Improvements
Frequencies			
No. of Freq. 5%	2	18	+ 16 Freq.
No. of Freq. 10%	13	23	+ 10 Freq.
Mean Freq. Ratio	1.121	1.038	+ 8.3 %
Standard Deviation	0.152	0.130	+ 2.2 %
Generalized Masses			
Ratio Global Masses	1.542	0.878	+ 42.0%
Ratio Mean Values	1.747	1.179	+ 56.8%
Standard Deviation	1.247	0.592	+ 36.5%
Kinetic Energies			
Ratio Global Energy	2.146	0.968	+111.4%
Ratio Mean Values	2.062	1.213	+ 84.9%
Standard Deviation	1.029	0.586	+ 44.3%
Cross-orthogonality			
Mean Value Diag. Term	0.8303	0.8108	- 1.9 %
Standard Deviation	0.1144	0.1286	- 1.4 %
Mean Non. Diag. Term	0.2588	0.2811	- 2.2 %
Standard Deviation	0.1815	0.1735	- 0.8 %

of withstanding the on-ground loads in a deployed configuration. The applicability of this approach has been verified on a space hardware solar array section of the Canadian Communication Satellite CTS.

Automatic Selection Procedure for Dynamic Degrees of Freedom (Figs. 10, 11)

AEO developed a theoretical approach which allows the selection of relevant degrees of freedom for dynamic tests in an automatic way and implemented the algorithm into the software package SIMP.

The selection is done neglecting rotational degrees of freedom, redundant rigidly coupled degrees of freedom and dof's which are insignificant for describing the desired mode shapes.

Under contract of the European Space Agency ESA, AEO demonstrated the efficiency of the approach using a mathematical model of the spacecraft dummy structure FLECS.

The method yielded an optimized number and distribution of the dof's to be measured in a dynamic test. Comparing the results with modal survey data obtained from ESA, it could be shown, that even experienced test engineers reach only about 60% or less of the significant and non-redundant degrees of freedom.

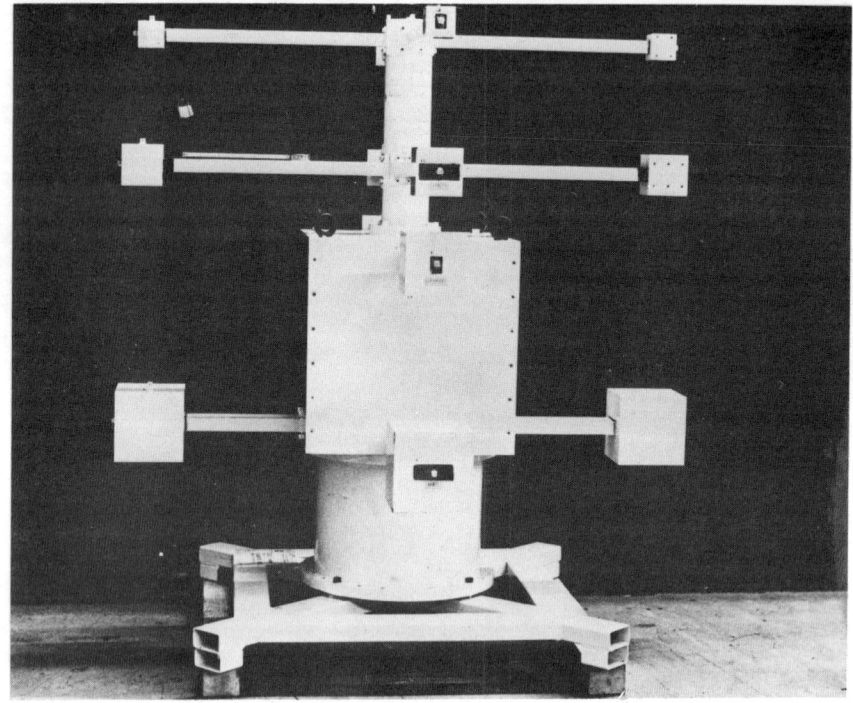

Fig. 7. The FLECS structure (courtesy of ESA/ESTEC) and its mathematical model.

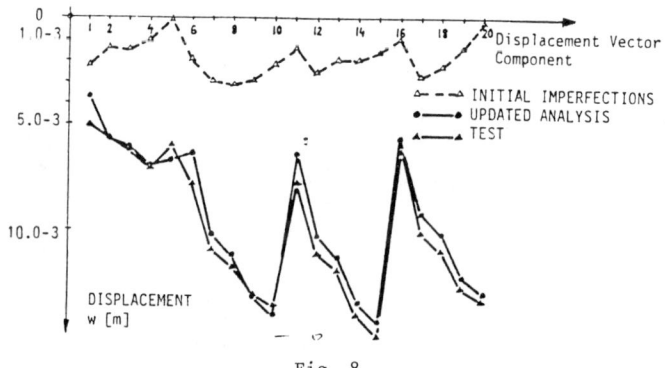

Fig. 8.

Solar Cell Analysis (Fig. 12)

AEO has integrated all the solar cell analysis performed in Europe's space projects into a unique software package. There is no other place in Europe where these analytical and computational tools are available in its entity. The work

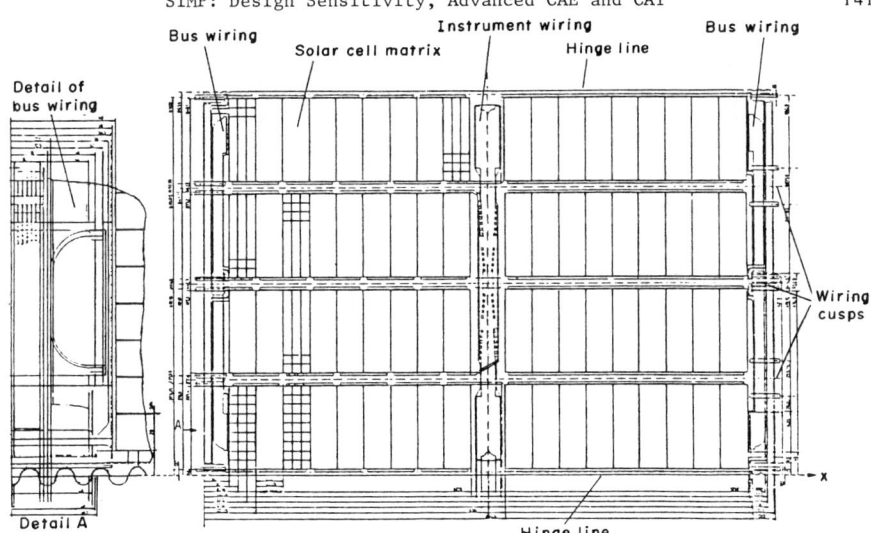

Fig. 9. The CTS solar array sub-panel assembly (SPA)

has been performed under contract of the European Space Agency, ESA. The purpose of it was to reduce cost of and time for the analysis by factors and also to reduce the workload of the Agency. The package deals with solar cell stresses, deformations, interconnector stresses and design optimization, lifetime analysis, wiring stresses, etc.

Other work was devoted to cassegrainian solar concentrators and terrestrial applications of solar cells. Together with the geometric nonlinear update of the CTS solar array, AEO has one of the most complete analysis tools for design analysis and design verification of solar arrays for space applications.

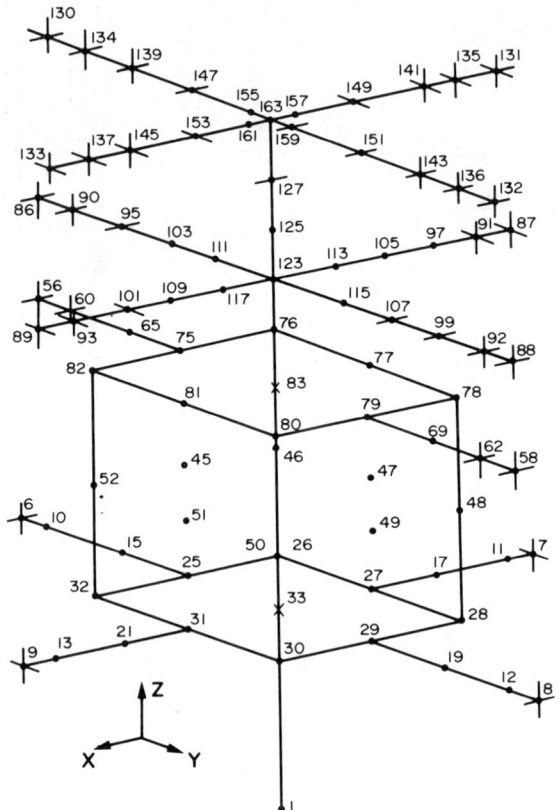

Fig. 10. Sets of relevant degrees of freedom.

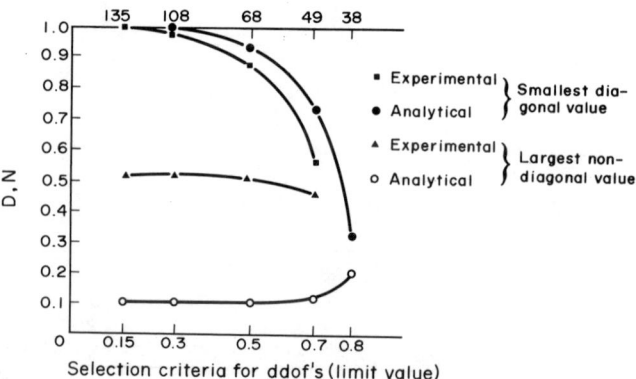

Fig. 11. Smallest diagonal (D) and largest non-diagonal (N) term of the cross-orthogonality for different sets of experimental and analytical determined degrees of freedom

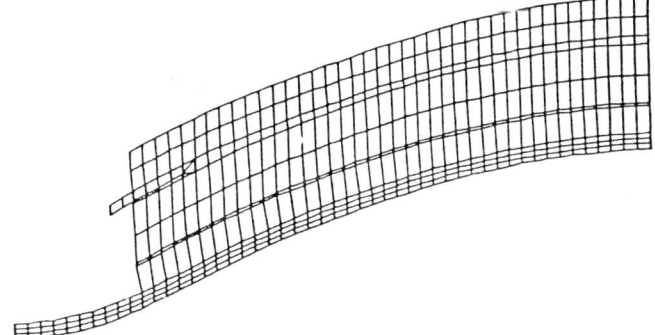

Fig. 12. Solar cell - 185, flexible structure.

STAR 2: STATIC ANALYSIS OF SPATIAL FRAME STRUCTURES

C. Katz-Fink-Stieda

Römerweg 1, D-8138 Andechs-Erling, Federal Republic of Germany

PROGRAM DESCRIPTION

Program STAR2 stands for the analysis of space frame structures including small displacement geometric and material nonlinearities.

The program is based on a displacement model. The structure has to be discretized in nodes, which can have rigid or elastic supports, and beam truss, cable or spring elements. In most cases only the real intersections of the elements have to be defined as nodes. Curves or folded structures are represented by several straight elements. Rigid elements are included.

Each beam has a straight centerline identical to the neutral axis but various cross sections and a separate shear center. Its properties are evaluated by an integration of the differential equations assuming along the axis piecewise constant stiffness and normal force. Shear deformation is an additional feature. Warping torque is not implemented.

The spring element takes care of nonlinear effects like a gap, yielding, friction or no-tension for forces and moments even between other structural elements.

A special cable-element is included.

Loading includes nodal forces and moments as well as unlimited loading of the beam elements anywhere. Temperature, prestressing, eccentric forces or displacements or special options for creep and geometric imperfections are available. Loading is conservative.

The modular package consists of the several programs which are connected by a database.

AQUA	Generation of cross sections (including shear, torsion and warping rigidity)
GENF	Generation of the structure
STAR2	Static analysis
AQUB	Design and material nonlinearities (including linear stress analysis for bending, shear, torsion and warping torque)
GRAF	Graphical display of results

MAXIMA Superposition of load cases

The design algorithms are rather general, but have special options for the German codes.

HANDLING

The program normally will run as batch job. The input is free format and has generation facilities. The input files can be created via a standard editor program or via a guided input with full screen options.

FIELD OF APPLICATION

The program STAR2 has its applications in the analysis of structural steelwork as well as reinforced concrete elements. It is especially suited for the analysis of structures near the buckling load.

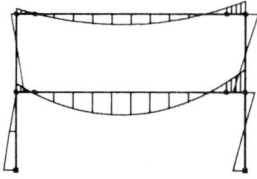

Example: Structural steelwork moments

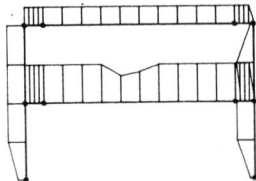

Example: Structural steelwork stiffness

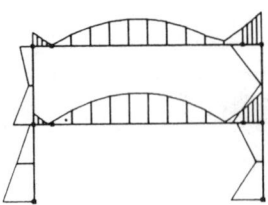

Example: Structural steelwork comparison stress

HARDWARE REQUIREMENTS

The program is written in FORTRAN-77 and will run on nearly any computer providing that language. Ready to run versions are sold for CP/M-80 and MS-DOS micro-computers. 2 MByte mass storage capacity (hard disk) is required. Graphics use the GKS standard functions.

STRUGEN: PRE- AND POSTPROCESSING WITH STRUGEN, COLOURED INTERPRETATION OF RESULTS OF STRESS OUTPUT

J. Simon-Weidner

IABG, Abteilung TDM, Einsteinstrasse 20, D-8012 Ottobrunn,
Federal Republic of Germany

ABSTRACT

In many areas of Finite Element Methods, isoparametric shape functions with e.g. linear or quadratic basis are well proved and known. Within these basic functions the local and undimensional coordinates are quite comfortable and transparent in use whe a detailed generation of a map of elements is necessary. For an overview of the structure (STRUGEN = STRUcture GENerator) it is possible to get information either interactively on a graphic terminal or on a plotter by hidden line technique. With the aim of postprocessing in an adequate comfortable way, a special hidden line technique for colour raster graphic terminals was developed. Most often, STRUGEN as pre- and postprocessor is used in connection with FEM-packages NASTRAN, ANSYS, STARDYNE, MARC and ADINA.

Used as postprocessor STRUGEN needs at least information about the geometry (elements and coordinates of nodes) and results from a FEM-calculation as data per nodal point, which represents the coloured output. To display the deformed structure, the displacements of all nodes are necessary, title and text of results can be defined separately. If the structure is very large and the computation of the hidden line drawing needs some time, it is efficient to use STRUGEN as postprocessor in the batch mode writing the whole coloured output as an image on a tape. Later in an interactive session it is possible to read this tape without high computation time; all information for drawing the coloured result   saved on this tape. STRUGEN is most often used as a postprocessor for the interpretation of equivalent stresses, temperatures and amplitudes from modal analyses.

PREPROCESSING

According to details in technical designs a generation of nodal points may be performed in local coordinate systems. For the special case of intersection points, which are common to two regular surfaces, the exact location is available by the use of connected coordinate systems.

In dividing a structure into elements, the concept of master points and master elements means that only some characteristic points are needed for the generation. Each master point may be given in any local coordinate system (cartesian, cylindrical or spherical). A transformation to the global system will be made, which is common to all points. A later transformation of previously defined master points to any other coordinate system is possible.

The master elements are based on the well-known isoparametric shape functions with dimensionless local coordinates. According to the generation index in a local (dimensionless) direction of a master element, the local coordinate will be subdivided into intervals yielding new elements. The types of elements are bars, plates and shells and volumes. These elements are defined with and without midside nodes.

A generation of segments by definition of geometric superelements is realized. Geometrical additions can be performed by transposition or reflection of the defined superelement consisting of different types of elements. A 3-D-structure may also be generated from a given 2-D-structure by a special option of superelement technique.

POSTPROCESSING

For a visualization of the structure, computed plots on paper or interactive graphics on terminals are available. Three different methods of hidden line techniques may be used. These are:

- the line area method
- the line raster technique and
- the area raster method.

The line area method (black/white drawing) checks each line as a part of an element against all other surfaces with the aim of visibility. Some artificial techniques for minimizing the computer memory requirement and tape operations are necessary.

The line raster technique (coloured graphic) is based on the ideas that lines next to the observer push away all other information within the local range of a raster line.

The area raster method (coloured graphics) draws the areas of the elements from the background to the foreground. Each drawn area will fill up with a colour. The areas of the foreground cover all the other element faces drawn just before.

The coloured interpretation of a result can be performed by the use of a direct access tape with the nodal value of interest written at the record corresponding to the node number. The result according to an index of colour will be coded to the third dimension coordinate of a node (perpendicular to the drawing plane). In this way, during the calculation of visibility the information of colour ($\hat{=}$ result) is presented simultaneously. The interpolation of colour can be easily performed using either the line-raster or the area-raster method.

In some cases a hidden line interpolation regarding the deformed structure is wanted. The scale of deformation may be prescribed by the user. If the scaling factor will be computed by the program, a value will be chosen so that the deformed drawing of the structure would change the needed range of area by 5% compared with the needed area of the undeformed structure.

Text strings assigning the result and the plot title are declarable separately.

During the postprocessing different operating possibilities may be used. The output may be on paper (plot) or a graphical terminal. The kinds of illustrations are transparent or hidden line technique. The drawing options are black/white (paper, terminal) or coloured (RAMTEK terminal), the performance using hidden line calculations may be made interactive or in the batch mode. The tape saved with all the needed information may be used later as the input for an interactive

STRUGEN: Pre- and Postprocessing with STRUGEN   149

coloured illustration on raster graphic terminal (RAMTEK).

INTERFACE TO AND FROM DIFFERENT FEM-PACKAGES

A structure generated by STRUGEN with all the data of nodes and elements may be used as input for different FEM-programs. By reason of their flexibility, the interfaces (programs for conversion) exist separately from the main program. A conversion exists for example from the STRUGEN generated data to NASTRAN, ANSYS, STARDYNE and ADINA.

A second conversion program exists for reading NASTRAN, ANSYS, MARC, or STARDYNE data into STRUGEN. If any of the FEM-packages are revised, the necessary adaption in the interface will be quickly completed.

EXAMPLE FOR POSTPROCESSING (FIGS 1, 2)

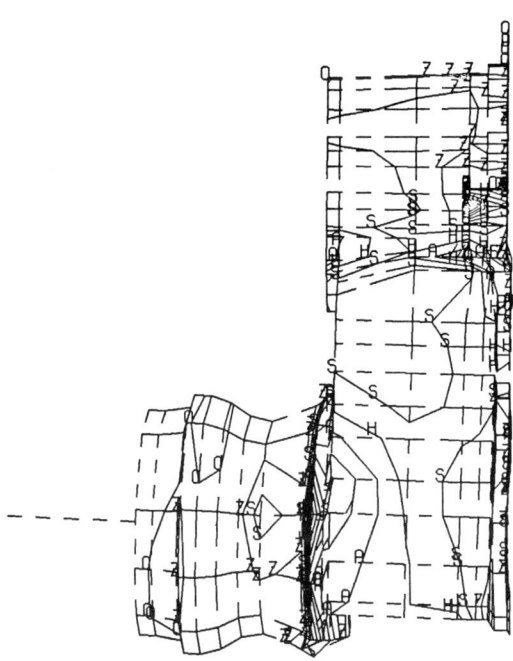

VIEW POINT   -262.00000
               0.00000
              -965.00000

STRUGEN   GENERIERUNG + AUSWERTUNG * N A S T R A N ANALYSE * ACHSAUFNAHME 01/03/84

MISES SPANNUNG   O +100.-01 Z +255.-01 S +411.-01 H +566.-01 A +721.-01 / +876.-01 ≤ +103.+00

DISP.FAC   7.59   + +118.+00 % +134.+00 ∧ +149.+00 ✻ +165.+00 $ +180.+00 ≡ +196.+00 X +211.+00

Fig. 1. Hidden-line postprocessing of a deformed structure with automatically scaled deformation factor. Some equivalent stresses (v. Mises) are shown. (By kind permission of company Linde, Aschaffenburg).

- Part of a hydraulic engine suspended by wheels
- civil engineering
- static
- solids, sheels and beams
- total number of degrees of freedom $\sim$ 4000
- analyses made by NASTRAN, pre- and postprocessing made by STRUGEN
- postprocessed region contains 545 elements with 615 nodes, CPU-time for postprocessing by STRUGEN needs 477 sec. for black/white drawing on a CYBER 855.

Potential users are companies in civil engineering, aeronautics etc.

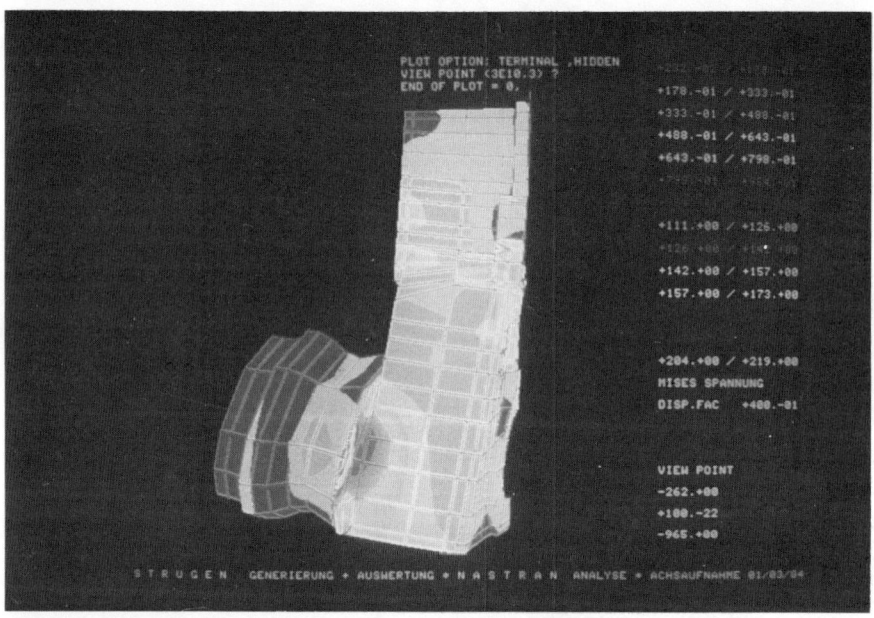

Fig. 2. Simon-Weidner (Niku Lari)

Y12M: SOLVING LARGE SYSTEMS OF LINEAR ALGEBRAIC EQUATIONS BY THE USE OF PACKAGE Y12M

Z. Zlatev

*Air Pollution Laboratory, Danish Agency of Environmental Protection,
Risø National Laboratory, DK-4000 Roskilde, Denmark*

ABSTRACT

The solution of large systems of linear algebraic equations is as a rule an essential part of the global computational process when structural analysis problems are treated numerically. Package Y12M contains several subroutines by the use of which some classes of large systems of linear algebraic equations can be solved in an efficient way. Therefore subroutines of package Y12M could be attached to existing programs for handling of structural analysis problems. Subroutines of this package could also be considered (as solvers of large systems of linear algebraic equations) in the development of new programs for structural analysis problems. The main properties of the subroutines of package Y12M, which make them attractive as solvers of systems of linear algebraic equations in large programs are:

(a) *the easy data input,*

(b) *the performance of significant arithmetic operations only (all operations including zero elements, or even small non-zero elements in some options, are omitted),*

(c) *the minimization of the storage requirements,*

(d) *the evaluation of the errors made (in the solution of the system of linear algebraic equations)*

(e) *the attempt to improve (optionally) the accuracy of the solution found.*

These properties of the subroutines of package Y12M are discussed in the following sections.

THEORETICAL BACKGROUND

Let $A \in \mathbf{R}^{n \times n}$ and $b \in \mathbf{R}^{n \times 1}$ be given. Consider the system of linear algebraic equations:

$$Ax = b \quad (x \in \mathbf{R}^{n \times 1}, \quad rank(A) = n) \tag{1}$$

Systems of this type appear often in the numerical handling of many problems in engineering and science. Moreover, very often the order n of matrix A is large and this causes difficulties both in regard to the computing time needed and in regard to the storage occupied by the data in the computer memory. Such difficulties may arise even when big modern computers are used. Therefore it is important to reduce both the computing time needed in the solution of (1) when n is large (by carrying out only the arithmetic operations involving non-zero elements of matrix A) and the storage occupied in the computer memory (by storing the non-zero elements of matrix A only). These two goals are achieved when some sparse matrix technique is applied. Therefore sparse matrix technique algorithms are implemented in all subroutines of package Y12M[3,5,8,10]. However, the old-fashioned use of sparse matrix technique only may be insufficient in some situations. In order to illustrate this statement let us assume that the classical Gaussian elimination is used to calculate two triangular matrices L and U such that

$$LU = A \quad (L \in \mathbf{R}^{n \times n}, \quad U \in \mathbf{R}^{n \times n}) \tag{2}$$

The calculations needed to obtain the matrices L and U are carried out by the formula:

$$a_{ij}^{(k+1)} = a_{ij}^{(k)} - a_{ik}^{(k)} a_{kj}^{(k)} / a_{kk}^{(k)} \quad \text{with} \quad a_{ij} \in A, \quad a_{ij}^{(1)} = a_{ij}, \quad a_{kk}^{(k)} \neq 0,$$

$$k = 1, 2, \ldots, n-1; \quad i = k+1, k+2, \ldots, n; \quad j = k+1, k+2, \ldots, n. \tag{3}$$

If $a_{ij}^{(k)} = 0$ but neither $a_{ik}^{(k)} = 0$ nor $a_{kj}^{(k)} = 0$, then $a_{ij}^{(k+1)} \neq 0$

and, thus, a new non-zero element, fill-in, is created in position (i,j) at stage k of the Gaussian elimination. It is quite obvious that the process of the Gaussian transformations is not very efficient when many fill-ins are created. Experience (see, for example,[1-5,9,10]) shows that, unfortunately, this does happen when sparse matrix techniques only are in use (even when special pivotal strategies are applied;[1-6,10]). Therefore an additional device is implemented in the subroutines of package Y12M. This device can optionally be used and if it is used, then all non-zero elements of matrix A which in the course of the Gaussian elimination become smaller in absolute value than a prescribed parameter, the drop-tolerance, are effectively considered as zero elements. This means that such elements are removed from the storage occupied in the computer memory and that no arithmetic operations with such elements are performed in the further calculations. Thus, both the storage and the computing time are normally reduced,[3,5,7-10] in this case. The reduction could be very great when positive values of the drop-tolerance are specified. Of course, the factors L and U calculated by the use of a positive drop-tolerance are not very accurate and, therefore, an iterative refinement process,

$$r_i = b - Ax_i, \quad d_i = (LU)^{-1} r_i, \quad x_{i+1} = x_i + d_i \quad (x_0 = (LU)^{-1} b, \quad i=1,2,\ldots,p)$$

$$\tag{4}$$

is carried out (in package Y12M when a positive drop-tolerance is chosen) in order to regain the accuracy lost during the Gaussian elimination. Thus, the method implemented in package Y12M is a compromise between the direct methods and the iterative methods and in[2-10] it is shown, by many numerical examples arising in different fields of science and engineering (as, for example, nuclear spectroscopic resonance theory, heat conduction, least squares problems), that for some classes of matrices this mixture of direct and iterative methods performs very well.

Y12M: Solving Large Systems of Linear Algebraic Equations   153

DESCRIPTION OF THE PACKAGE

The package contains several subroutines. The subroutines are written in FORTRAN. Both single precision versions and double precision versions of the subroutines are available. Only standard FORTRAN 66 statements are used. This ensures portability of the package. The subroutines have been run on several big computers (see, for example,[8]). Further details about the package and about the different tests carried out with the package can be found in[8-10]. It must be mentioned here that special versions, in which some UNIVAC facilities are applied to reduce further the storage needed, have been developed[10].

The data input is very simple. The user must store the values of the non-zero elements of matrix A in a one-dimensional REAL array, say A1, of length NN (where NN should be larger than NZ; NZ being the number of non-zero elements in matrix A). The order in which the non-zero elements of matrix A are stored in array A1 can be arbitrary. The only requirement to the user (concerning the input data) is the following. If a non-zero element of matrix A is stored in A1(K) ($1 \leq K \leq NZ$), then its row number must be stored in RNR(K) and its column number must be stored in SNR(K) (RNR and SNR being INTEGER arrays of lengths NN and NN1, where NN is as above, while NN1 should be larger than NZ but in general is smaller than NN). This is perhaps the simplest way to organize the input operations. Note that the user could store the non-zero elements in any order and, thus, choose the order which is most convenient for him. The package will accept any ordering of the non-zero elements. This is very useful when finite elements are applied and, especially, when some automatic mesh generation is implemented in the program which calls the subroutines of package Y12M.

The user has a possibility to obtain some information about the accuracy of the solution (together with the solution itself). This possibility could be exploited to develop some acceptability criteria or to develop some other interface devices when the package is used (as a solver for systems of linear algebraic equations) in a large program.

HARDWARE COMPATIBILITIES

It is assumed that matrix A can be held in core when the subroutines of package Y12M are used. Therefore the package should be used on big computers. If this is the case then no peripherals are needed (excepting the UNIVAC-oriented version mentioned in the previous section). The subroutines of package Y12M are in the standard library at RECKU (the Regional Computing Centre at the University of Copenhagen). These subroutines, together with a full documentation of the package and with several demonstration programs, are available on magnetic tape. Requests should be addressed to J. Wasniewski, RECKU, Vermundsgade 5, DK -2100 Copenhagen, DENMARK.

EXAMPLES OF APPLICATIONS

Package Y12M has been used in several different areas; see[7-10]. The example discussed below is not a structural analysis problem but, nevertheless, the assemblage of the finite elements used in the space discretization of many structural analysis problems leads to the solution of similar systems of linear algebraic equations. The problem arose in modelling of heat accumulation in underground water-tanks by solar batteries. The governing equation is (the unknown function T being the temperature):

$$-\frac{\partial T}{\partial t} + \frac{\partial^2 T}{\partial x^2} + \frac{\partial^2 T}{\partial y^2} = 0 \tag{5}$$

where $T(x,y,0)$ is a given initial distribution. The boundary conditions are given on Fig. 1. The number of finite elements used in the space discretization of the domain shown on Fig. 1 is 840, the number of nodes is 470. Thus, a system of 470 linear algebraic equations is to be solved at each time-step. The number of time-steps carried out is 1010 (resulting in integrating (5) over a time interval of 5 years). The integration begins with a time-stepsize of 6 hours and the time-stepsize is gradually increased to 96 hours. An attempt to use the old factorization (the old factors L and U) of matrix A is very successful in this case because matrix A changes only when the time-stepsize is changed. As mentioned above the number of time-steps is 1010 but a new factorization of matrix A is calculated only 12 times. The problem is solved directly (using the LU factorization calculated by accepting *all* fill-ins and without any attempt to apply (4) and by the iterative refinement process (4) removing all fill-ins smaller in absolute value than 0.001. Some results for a typical time-step are given in Table 1. It is seen that the use of the iterative refinement process combined with removing small non-zero elements is very efficient for this problem (giving a considerable reduction both in computing time and in storage).

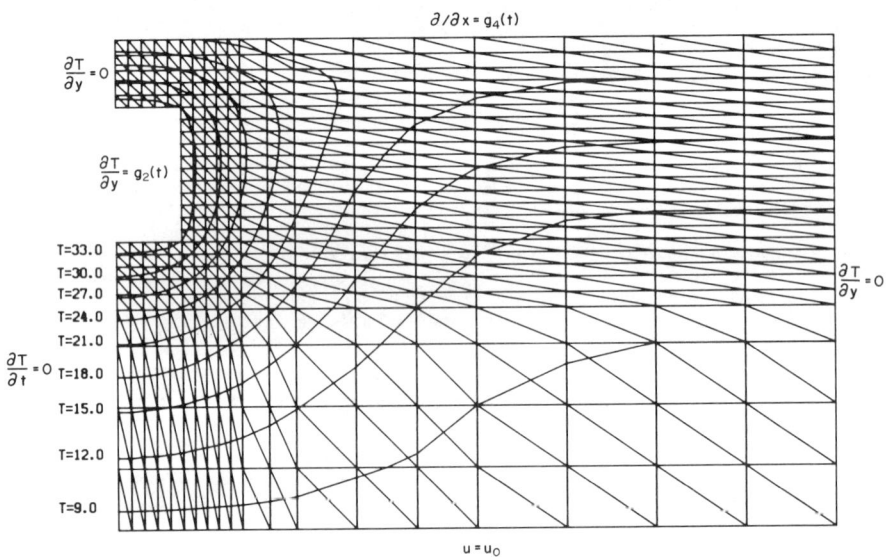

Fig. 1. The finite element discretization of the domain of (5) with niveau-curves of the solution obtained.

TABLE 1 Some characteristics obtained at a typical time-step in the numerical integration of (5). NZ is the number of non-zero elements in A before the beginning of the Gaussian elimination. NZ+NF is the number of non-zero elements in the factors L and U. The computing time is measured in seconds. The run is performed on an IBM 3081U with the FORTH compiler with OPT=02.

Mode of solution	NZ	NZ+NF	Computing time
Direct solution	3089	9741	1.53
Iterative refinement	3089	5229	0.64

Some other examples including

(i) problems arising in nuclear spectroscopic resonance theory

and

(ii) linear least-squares problems

are discussed in[7-10].

It should be emphasized here that the use of package Y12M is not the best choice in all cases. If the problem is such that A is a banded matrix with a narrow bandwidth, then the application of another solver (based on the use of band-matrix techniques, frontal or multifrontal methods etc.) may be more profitable. If matrix A is such that some purely iterative procedures (as, for example, the SOR algorithm) is quickly convergent, then the use of such a procedure should be recommended. This shows that in a good program for treatment of structural analysis problems it is worthwhile to implement several different solvers for systems of linear algebraic equations in order to ensure a great degree of flexibility of the program and to increase the efficiency for a wide class of problems (provided that the proper solver is selected).

REFERENCES

1. Zlatev, Z. (1980). On some pivotal strategies in Gaussian elimination by sparse technique. *SIAM J. Numer. Anal.*, 17, 18-30.

2. Zlatev, Z. (1980). On Solving Some Large Linear Problems by Direct Methods, DAIMI PB-111. Department of Computer Science, University of Aarhus, Denmark.

3. Zlatev, Z. (1982). Use of iterative refinement in the solution of sparse linear systems. *SIAM J. Numer. Anal.*, 19, 381-399.

4. Zlatev, Z. (1982). Comparison of two pivotal strategies in sparse plane rotations. *Comput. Math. Applics.*, 8, 119-135.

5. Zlatev, Z. (1985). Sparse matrix technique for general matrices with real elements: Pivotal strategies, decompositions and applications in ODE software. In: D. J. Evans (ed.) *Sparsity and its Applications*, Cambridge University Press, Cambridge, pp. 185-227.

6. Zlatev, Z. (1985). A general scheme for solving large algebraic problems. *IMAGS J. Nimer. Math.*, 1, pp. 177-186,

7. Zlatev, Z. and P. G. Thomsen (1981). Sparse matrices - efficient decomposition and applications. In: I. S. Duff (ed.) *Sparse Matrices and Their Uses*. Academic Press, London, pp. 367-375.

8. Zlatev, Z., J. Wasniewski and K. Schaumburg (1981). Y12M - Solution of Large and Sparse Systems of Linear Algebraic Equations. Springer, Berlin.

9. Zlatev, Z., J. Wasniewski and K. Schaumburg (1982). Comparison of two algorithms for solving large linear systems. *SIAM J. Sci. Statist. Comput.*, 3, 486-501.

10. Østerby, O. and Z. Zlatev (1983). Direct Methods for Sparse Matrices. Springer, Berlin.

FINITE ELEMENT SOFTWARE FOR STRESS ANALYSIS OF LAMINATED COMPOSITES

J. Mackerle

*Linköping Institute of Technology, Department of Mechanical Engineering,
S-581 83 Linköping, Sweden*

ABSTRACT

In recent years composite materials, especially fiber-reinforced laminates, have found many applications in engineering structures. This paper gives a list and the description of some general purpose and special purpose finite element programs which are used for stress analysis of these materials. For general purpose programs the description given is limited to capabilities which are specific for the analysis of composites; special purpose programs are presented in more details, both in a descriptive and in a tabular form respectively.

A comprehensive bibliography of published papers contains references dealing with application of the finite element method for the analysis of composite structures. Also theoretical papers are included.

INTRODUCTION

The industry has utilized fibre-reinforced composite materials for over a decade. Most composites are used to improve mechanical properties such as strength, stiffness, toughness and performances at elevated temperatures. In many cases the composites are used to reduce material costs. Total weight of the structure is often a factor in the design.

Composite materials can be divided into fiber-reinforced composites and particle-reinforced composites. Fibrous composites, discussed in this paper, can be classified as single-layered or multilayered (angle-ply) composites. Most composites in practical applications are multilayered. It means that they consist of several layers of single-layer composites but with different fiber orientation. These layers, usually very thin, are bonded together.

Continuous fiber-reinforced composites consist of long fibers embedded in the matrix, discontinuous fiber-reinforced composites contain short fibers in the matrix respectively. All fibers aligned in a single-layer in only one direction form unidirectional composites, with a second complement direction form bidirectional (cross-ply) composites. Hybrid laminates are multilayered composites that consist of layers of different fibers and/or matrix material.

With the rising use of composite materials there is a great need for understanding their behaviours under specific conditions. This resulted in a number of special purpose finite element codes developed during the last decade and in the inclusion of new options/elements into the widely used general purpose programs.

Some of the programs directed for stress analysis of laminated composites are listed and briefly described in the next sections.

PROGRAM REVIEW

Finite element programs described in the following sections are divided into two main categories:

(1) - general purpose programs
(2) - special purpose programs

Almost all presented general purpose programs are well known. Therefore description of their capabilities is directed to the applications in connection to the analysis of composite materials/structures. For more detailed general description the reader is referred to references 1-3 or to the respective program documentation. Also review paper on the evaluation of software packages for stress analysis of laminated composites is recommended in reference 4.

The main emphasis of this review paper is the presentation of special purpose programs for stress analysis of composites. To the author's knowledge no extensive presentation is available in the "open" literature.

Three categories of software according to the user's point of view are represented in the survey:

- programs serving primarily as a research tool, developed to evaluate new formulations, elements, etc, not intended for other users; documentation available is very often poor
- programs developed as a research tool with limited documentation, available for a nominal fee for other users for non-commercial applications
- programs commercially available, fully documented

The originators of these programs are many, their motivation and qualification varies. Programs chosen for this paper have been developed within the universities, at different research institutions and at industrial companies.

The subject is too large to be extensively covered in a single paper, therefore only a brief description of software is outlined. The descriptive survey gives information about the program developer, source for detailed program information, and program category. Thereafter follows a short program description, element library available, and other special program capabilities. Information about hardware, program availability, specific program applications are also given. Some references are given at the end of each program description. These are oriented to the program documentation, specific applications of program-developer research activities. References are listed in Appendix under respective subject category. Range of the program applications is divided into area of:

- macromechanical analysis
- micromechanical analysis
- fracture mechanics
- joints/bonding
- other areas of application

It is not our intention to evaluate presented programs but first of all outline in some detail what type of software there is on the "market" for a potential user for the stress analysis of composites.

GENERAL PURPOSE FINITE ELEMENT PROGRAMS FOR THE ANALYSIS OF COMPOSITE MATERIALS

The following programs are described in this section: ABAQUS, ADINA, ANSYS, ARGUS, ASAS, ASKA, DIAL, MARC, MSC/NASTRAN, NISA, PAFEC, SAMCEF, SAP versions, SPAR, and WECAN.

Program Name: *ABAQUS*

Source for program information: Hibbitt, Karlsson & Sorensen, Inc., 35 South Angell Street, Providence, Rhode Island 02906, USA. Also available through Control Data Cybernet services.

Element library: 4 node or 8 node reduced integration general layered shells. 2 or 3 node axisymmetric layered shells. Either numerical integration or a user-supplied section stiffness matrix. Any material model can be used with any element type. No limit on the number of different materials or elements in a model. Anisotropic option is available for all elements. Thermal loads are permitted.

Material library: Elastic with temperature dependence, isotropic, orthotropic or anisotropic definitions. Elastic-plastic with Hill's anisotropic hydrostatically independent yield. Creep- isotropic or anisotropic properties are permitted.

Other program capabilities: Multiple coordinate system input. Cartesian, cylindrical, spherical with any point of origin. Simple keyword, free-format input. Versatile restart.

Hardware: CDC, IBM, VAX, Univac, Prime.

Program Name: *ADINA*

Source for program information: K. J. Bathe, Massachusetts Institute of Technology, Dept. of Mechanical Engng. (3-356), Cambridge, MA 02139, USA. Available through ADINA Engineering, Munkgatan 20D, S-722 12 Västarås, Sweden or ADINA Engineering Inc., 71 Elton Ave., Watertown, MA 02172, USA.

Element library: For 2D problems- 3-8 node isoparametric plane stress/strain and axisymmetric elements. For 3D problems- 8-21 node isoparametric solid and thick shell elements. Elements may also be used for the geometrically nonlinear analysis. Multilayered elements are not available. User-supplied elements are accepted.

Material library: Orthotropic linear elastic. Shear coupling.

Other program capabilities: Substructuring in static and dynamic analysis. Restart from initial conditions. ADINA-IN is a preprocessor program, ADINA-PLOT is a postprocessor program available.

Hardware: CDC, IBM, Univac, VAX, Prime, Burroughs, Cray, Cyber.

Program Name: *ANSYS*

Source for program information: Swanson Analysis Systems Inc., P.O. Box 65, Houston, PA 15342, USA.

Element library: Elements with linear orthotropic material properties- 2D plate elements, axisymmetric conical shell, 3D membrane element, 2D axisymmetric ring element, 3D isoparametric solid elements, 3D solid element, axisymmetric quadrilateral element with non-axisymmetric loading, thick-shell isoparametric element. Laminated layered shell element with up to 15 layers and in-plane or normal loads. Quadrilateral shell element-interlaminar shear. 3D crack-tip orthotropic element. User-supplied elements are accepted. 2D and 3D interface element, combination element. A compound structure may be modelled in terms of several super elements, each super element itself being an assembly of ANSYS elements or other super elements.

Material library: Orthotropic material properties may be used with all plane and solid elastic structural elements and with all heat transfer elements. An elastic structural element allows user input of a complete material matrix for general anisotropic behaviour. All elastic material properties may be functions of temperature.

Other program capabilities: Cartesian, cylindrical or spherical coordinates are available. Various user-defined local coordinate systems may also be used. Nodal coordinate systems may be rotated when it is desirable to input and output displacements and force components in directions other than the global coordinate directions. Some elements have their own element coordinate system for inputting orthotropic material properties and for interpreting output stresses. Element stiffness matrices, once formed, are saved and re-used throughout the run where necessary. Different pre- and postprocessors are available. User may define own failure criterion by using arithmetic operations and parametric input in POST1.

Hardware: CDC, IBM, Univac, Amdahl, Burroughs, Cyber, Fujitsu, Data General, Harris, Honeywell, Prime, VAX, Cray.

Program Name: ARGUS

Source for program information: Merlin Technologies, 977 Town and Country Village, San Jose, California 95128, USA. Available through the Control Data Co Cybernet service.

Element library: It contains 4 or 5 node quadrilateral and 3 node triangular elastic thin plate/shell element with anisotropic multilayered construction. This element can have any number of orthotropic layers and the material axes of each layer can have any orientation with respect to the element local coordinates. Element layers can be made of the same or different materials, same or different layer thickness. The physical and geometrical properties of the layers may be non-symmetric with respect to the element mid-surface. Doubly curved layered 8 node shell anisotropic element is also available. Thick shells and 3D solid elements are available for orthotropic materials (monocoque).

Material library: Orthotropic elasticity and composite materials. Temperature-dependent properties are accepted.

Other program capabilities: Cartesian, cylindrical and spherical coordinates. Skewed coordinates and DOF. Multi-axial failure criteria for composites are included. Biaxial maximum stress failure criterion for inplane normal stress, and maximum shear stress failure criterion for inplane/interlaminar shear stress. Output-element layer stresses are given in terms of layer material axes on the stress printout and in terms of element local coordinates on post-plots. Four distinct levels at which job restart may be initiated. Different pre- and postprocessors are included.

Hardware: CDC, Univac.

Program Name: ASAS

Source for program information: Atkins Research and Development, Woodcote Grove, Ashley Road, Epsom, Surrey KT18 5BW, England.

Element library: 8 node quadrilateral curved thick shell element (Ahmad, isoparametric), 6 node triangular curved thick shell element (Ahmad, isoparametric). Multilayered option. Several other element types can have orthotropic material properties. User-specified elements are accepted.

Material library: Laminate design is restricted to thermo-elastic material, but each lamina can be generally anisotropic. Any number of layers is allowed. Layers may be non-symmetric with respect to the element mid-surface. The composite capability is limited to linear elastic static and dynamic analysis. Max stress, max strain, Tsai-Hill and Tsai-Wu failure criteria are included in the program.

Other program capabilities: Free-field input. Global/local coordinate systems. Multilevel substructuring. Restart option. Several pre- and postprocessors are available.

Hardware: CDC, IBM, Univac, Honeywell, DEC, Prime, Xerox, Sigma, ICL, VAX, Cray, Apollo.

Program Name: ASKA

Source for program information: IKOSS, IKO-Software Services GmbH, Albstadtweg 10, D-7000 Stuttgart 80, West Germany.

Element library: Many 2D and 3D element types can be used in connection with the anisotropic material properties. Triangular plane shell layered element including shear, quadrilateral sandwich element including shear. Triangular plate layered element. User-supplied elements are accepted.

Material library: Isotropic and anisotropic material behaviour. An interactive postprocessor evaluates ASKA results for anisotropic surface elements. The margin of safety against laminate failure is calculated using the max strain criterion. Either the most strained ply or all plyes in a layup may be printed.

Other program capabilities: Global/local coordinate systems may be defined. Free-field input. Various stress output, user-defined. Multilevel substructuring technique. Restart at any stage of computation. Access to the internal data structure. FEPS is the graphics package for ASKA.

Hardware: CDC, IBM, Univac, Cyber, ICL, Burroughs, Honeywell, Prime, Amdahl, Robotron, Cray, VAX.

Program Name: *DIAL*

Source for program information: Lockheed Missiles and Space Co., Manager of Structural Dept., P.O. Box 504, Sunnyvale, CA 94086, USA.

Element library: The isoparametric elements can be of mixed order. The shell and curved beams are degenerated from the solid isoparametric elements and allow through the thickness shear deformation. Layered shells and plates are available. Several elements have an option for the orthotropic material properties. Curved thick shells 2D and 3D, hybrid stress flat thin shell, curved isoparametric 3D solid elements. User-supplied elements are accepted.

Material library: Material properties are specified in the MATL processor. Types of materials available for composites- isotropic and general anisotropic- linear, nonlinear elastic orthotropic- different tension and compression properties (Jones-Nelson). All materials can be temperature-dependent. Complete direct stress recovery in laminated composites and computation of margins of safety using any of five failure criterion-Tsai-Wu, Hoffman, Tsai-Hill, Maximum stress and Maximum strain.

Other program capabilities: Free-field input. Global and local coordinate systems can be defined. A standard Cartesian coordinate system is used but if desired a mesh can also be defined in any of several shell curvilinear coordinate systems. Multiple systems are permitted. Restart capability at almost any analysis step. Substructuring for the static and dynamic analysis. Several modules for data generation and plotting.

Hardware: Univac, CDC Cyber, VAX, Cray, Apollo, Prime, Harris.

Program Name: *MARC*

Source for program information: MARC Analysis Research Corporation, 260 Sheridan Ave., Suite 200, Palo Alto, California 94306, USA.

Element library: Beams with open or closed cross-sections with or without warping, axisymmetric or 3D shells, double curved shells, plates. All shell elements may be composed of an arbitrary number of layers, each of which may have orthotropic properties.

Material library: Anisotropic, time-independent behaviour is available for both elastic and elastic-plastic response (with different yield criteria or different hardening rules). Anisotropy of the yield function is included in association with deviatoric stress only.

Other program capabilities: Global and local coordinate systems. Cartesian, cylindrical, spherical coordinates. User subroutine for definition of own coordinate systems. Plot routines for conforming of layer quantities, deformed shapes, history plotting, variable vs variable. Multilayer plotting with shell or beam elements. Restart option is available. MENTAT is an interactive graphics pre- and postprocessor.

Hardware: IBM, CDC, Univac, Prime, Cray, Cyber, VAX, Data General, HP, Harris.

Program Name: *MSC/NASTRAN*

Source for program information: MacNeal-Schwendler Co., 7442 North Figueroa Str., Los Angeles, California 90041, USA.

Element library: Several element types have an option for the orthotropic material properties. Multilayered triangular and quadrilateral isoparametric plate and shell elements. Failure prediction element for fiber reinforced composites (QUAD4C). User-supplied elements are accepted.

Material library: Isotropic, orthotropic, anisotropic, temperature- and stress-dependent material properties. There are two methods of inputting material properties- the stiffness is inputted manually after calculating them separately, or entering each ply and letting NASTRAN calculate the stiffness. This second capability transforms the plate and shell elements into layered elements. Ply by ply stresses and ply failure criterias are availble as output. Hill, Hoffman and Tsai-Wu failure criterion is incorporated into the program.

Other program capabilities: Free-field input. Global/local coordinate systems. Cartesian, cylindrical or spherical coordinates may be used. Restart option. Multilevel substructuring. Post-buckled composite structures may be analysed. MSC/GRASP is an interactive pre- and postprocessor.

Hardware: IBM, Amdahl, Itel, Fujitsu, CDC, Cyber, Univac, VAX, Cray, Siemens.

Program Name: *NISA*

Source for program information: Engineering Mechanics Research Corp., P.O. Box 696, Troy, Michigan 48099, USA.

Element library: Plate/thick shell (quadrilateral and triangular) and solid elements (hexahedronal and pentahedronal) are available for modelling of complex composite structures. A family of quadrilateral and triangular general laminated shells with 6 to 24 nodes is also included in the element library. Shell elements are doubly curved. Fiber directions may be specified globally. Composite layers may be isotropic, transversely isotropic, orthotropic or anisotropic. Arbitrary (symmetric/unsymmetric) layup with multimaterials, any number of layers. Sandwich construction can be modelled. Composite elements are compatible with other element types. Elements include bending-extensional-twist-shear coupling even for flat plates. Smooth variation of lamination parameters (thickness, fiber angles, etc). Interlaminar shear effects are included. Exact distribution of interlaminar shear stress is found.

Material library: Temperature-dependent material properties can be specified for static, dynamic and heat transfer analysis. Isotropic, orthotropic and anisotropic (layered or sandwich composite) properties. Orthotropic material may be specified on the global level or the element level. The direction of orthotropy may be different for each element/node. NISA incorporates maximum stress, maximum strain, modified Hill and von Mises, Tsai and Wu, and delamination failure crteria. Delamination is based on interlaminar shear stress exceeding a user-specified allowable value.

Other program capabilities: Cartesian, cylindrical or spherical coordinate system. User-specified local orthogonal system. All coordinates are transformed to the global Cartesian system. Rotated coordinate system can also be established whenever it is desirable to input displacements and forces in directions other than the global coordinate directions. Plot routines are included in the program. Contour plots of layer stresses and stress resultants. Survey plot of critical stress components in all layers. Comparison of stresses for each layer with allowable stresses which may be different in tension and compression and may be temperature-dependent. Several pre- and postprocessors are available.

Hardware: NISA is machine independent. In present running on Cray, Amdahl, IBM, CDC, Univac, Honeywell, VAX, Prime, Harris, HP.

Program Name: *PAFEC*

Source for program information: PAFEC Ltd., Strelley Hall, Strelley, Nottingham NG8 6PE, England.

Element library: A group of anisotropic elements is available. A family of solid isoparametric hexahedronal elements, wedge elements, 5-12 node shell elements, membranes, plate bending elements. For a family of sandwich elements the compliance matrix may be specified. User-supplied elements are permitted. Orthotropic semiloof element with a multilayered facility.

Material library: Elements may be defined which have linear, nonlinear, anisotropic, laminated and temperature-dependent material properties.

Other program capabilities: Global/local coordinate systems. Cartesian, cylindrical and spherical systems are permitted. Multilevel substructuring technique is included in the program. PIGS is a general purpose interactive graphics pre- and postprocessor program.

Hardware: Prime, IBM, Cray, Itel, DEC, Honeywell, Burroughs, Iris, CDC, Univac, Perkin-Elmer, Xerox, GEC, ICL, SEL, Data General, Apollo, HP.

Program Name: *SAMCEF*

Source for program information: University of Liege, LTAS, Rue E Solvay, 21, B-4000 Liege, Belgium.

Element library: Quadrilateral and triangular isoparametric multilayer axisymmetric, plane stress and plane strain elements for bidimensional analysis, quadrilateral and triangular isoparametric multilayer axisymmetric element for asymmetric loading (Fourier analysis), 3D isoparametric multilayer shell and volume elements. User-supplied elements are accepted.

Material library: Linear and nonlinear user-defined material library. Specific options for analysis of composites; nonlinear stress-strain relations are included by the Sandhu model, Tsai-Hill, Tsai-Wu and Sandhu failure criterion, model for materials with different moduli in tension and compression, post-degradation analysis, interlaminar stress analysis. Static and dynamic, transient linear and nonlinear analysis, as well as steady state and transient linear and nonlinear heat transfer analysis may be handled. The material properties are introduced by an interactive preprocessor.

Other program capabilities: Local coordinate systems may be introduced. Multilevel substructuring technique is included. Restart option. SAMCEF can be used for optimization studies. Interactive programs for data generation and plotting are available.

Hardware: IBM, CDC, Univac, VAX, Siemens, Cray.

Program Name: *SAP* Versions

Source for program information: SAP V- SAP Users Group, Dept. of Civil Engng., University of Southern California, University Park, Los Angeles, CA 90007, USA. SAP 7- Structural Mechanics Computer Laboratory DRC 394, University of Southern California, Los Angeles, CA 90007, USA. SAP-80- E. L. Wilson, 1050 Leneve Place, El Cerrito, CA 94530, USA.

Element library: SAP V- Quadrilateral and triangular plane stress/strain, plate and shell orthotropic elements. Plate and shell general anisotropic elements. Variable-number-nodes thick shell and 3D isoparametric elements with orthotropic material properties. User-defined elements. SAP 7- 2D isoparametric plane stress/strain or axisymmetric, 3D isoparametric solid elements, 3-9 node plate elements with layered option, 4-16 node isoparametric curvilinear plate/shell elements. All elements can be orthotropic. 4-16 node general multilayer plate/shell element with any number of layers and variable thickness. Each layer must be isotropic or orthotropic. A nonlinear in-plane shear stress-strain relationship is permitted. Some elements can be used for geometrically nonlinear analysis. SAP-80- Beams and quadrilateral elements. Stacked design with different material properties in each layer is permitted but the program does not have orthotropic elements.

Material library: SAP V- Orthotropic and shear-coupled. Temperature-dependent material properties are permitted. SAP 7- Orthotropic and anisotropic model. Tsai-Hill and Tsai-Wu failure criterias are implemented. Layered sandwich material may be analysed. SAP-80- Only linear elastic isotropic material properties.

Other program capabilities: SAP V- Global/local coordinate system. Bandwidth minimization. Plot routines are available. SAP 7- Substructuring is implemented for static problems. Local coordinate systems may be defined. Free-format input. Restart capability. SAP-80- Global and local coordinate systems. Restart option. Limited model generation, generation of loading and boundary conditions. No plot routines. MODEL is an interactive preprocessor, PLOT is an interactive postprocessor available. Both programs can be used in connection with different SAP versions.

Hardware: SAP V- CDC, IBM, Univac, VAX, Burroughs. SAP 7- IBM, CDC, VAX, Prime, Data General. SAP-80- IBM PC, IBM PC's compatible, VAX, Cromemco, Radio Shack.

Finite Element Software for Stress Analysis 165

Program Name: *SPAR*

Source for program information: Engineering Information Systems (EISI), Inc., 5120 Campbell Ave., Suite 240, San Jose, California 95130, USA.

Element library: Triangular and quadrilateral membrane elements, triangular and quadrilateral plate bending elements, quadrilateral shear panel, tetra-, penta- and hexahedronal solid elements. All these elements may be anisotropic. Combined membrane/bending triangular and quadrilateral elements with anisotropic or layered anisotropic properties.

Material library: Linear elastic isotropic and anisotropic material properties.

Other program capabilities: Free-field input. Multiple coordinate systems. Substructuring technique is available. Restart after any processor function. Pre- and postprocessors are available.

Hardware: CDC, Univac, Prime, VAX, Cyber.

Program Name: *WECAN*

Source for program information: Westinghouse Computer Service, Advanced System Technology, 777 Penn Center Blvd., Pittsburgh, PA 15235, USA.

Element library: Several WECAN elements have an orthotropic and anisotropic material capabilities. 3D beams, 2D plane stress/strain elements, axisymmetric solids, plates and shells, and 3D isoparametric elements are available. User-supplied elements are accepted.

Material library: Linear elastic, orthotropic or anisotropic. Material library for some typical commercial available material systems such as E Glass/Epoxy, S Glass/Epoxy, Graphite/Epoxy and Kevlar/Epoxy are available through a preprocessor called IPAC. The IPAC code also allows the user to input the experimental data. Temperature dependence of material properties is permitted. The safety margin analysis can be performed. Three failure criteria are available in the postprocessor IPAC. These criteria are - Tsai-Wu failure criterion, max stress and max strain criteria.

Other program capabilities: COMPOSITE is a part of WAPPP program to calculate the effective moduli for macromechanics analysis of multilayered fibrous composite materials. The program is limited to elastic stress analysis. The WAPPP is the WECAN Auxiliary pre- and postprocessor program package. Global and local coordinate systems in Cartesian, cylindrical, spherical and user's own MAPING FORTRAN subroutine are provided. WECAN contains two different methods of substructure analysis. Plot programs are available. The stress contour plots showing discontinuity between material boundaries can be provided. WECAN can be used for the static, dynamic, heat transfer and moisture content analysis of composite structures.

Hardware: WECAN and WAPPP on CDC and Cray, FIGURES (interactive graphics pre- and postprocessor) on Prime, VAX, Apollo.

SPECIAL PURPOSE FINITE ELEMENT PROGRAMS FOR ANALYSIS OF COMPOSITE MATERIALS

The following programs are described in this section: AB5U, APOC, BEAM, BEOS, BOLT, CCMFE, CLFE2D, COMPASS, COMPOSITE/Japan, COMPOSITE/USA, COSMOS, EPT3, FE2D and FE3D, FIBER, FRACOM, FRAMEAP, GAMNAS, LAMINA, LAMPS, LASSAQ, LAVIB, LSHELL and NLSHELL, MULTPN, NALCOM, NAPLES and VAPLES, NOLIN, NONCOM and NOLACS, PAC78, PLASCAN and PLASTOSHELL, Q3DG, SANWAT, SAPCOM, SDSAP-CJ, STAEBL, SUN/FL, THICK, THVROLL, TSJ1, WANG, and WAVE PROPAGATION.

Tables I and II give a "quick-review" of program capabilities according to the

TABLE 1

TAB 1

Program name	2D analysis	3D analysis	Macromech static	Macromech dynamic	Micromech static	Micromech dynamic	Heat transfer	Buckling	Fracture mechanics	Joints, bonding	Hardware
AB5U	●		●								Not known
APOC	●	●	●								Felix
BEAM	●	●	●					●			Cyber
BEOS	●		●					●			IBM, Univac, VAX
BOLT	●								●	●	VAX, IBM, Amdahl
CCMFE		●	●		●				●	●	VAX, IBM
CLFE2D	●		●						●		IBM
COMPASS	●		●N								CDC
COMPOSITE/Japan		●	●N					●	●		IBM, Hitac
COMPOSITE/USA		●	●						●		CDC Star
COSMOS [1]	●		●	●							IBM
EPT3 [2]			●								IBM
FE2D, FE3D	●	●	●N		●	●					IBM
FIBER		●			●N						Univac
FRACOM	●								●		Nord, VAX
FRAMEAP	●	●	●	●			●	●	●N	●	CDC, IBM, VAX
GAMNAS	●				●N				●	●	Prime
LAMINA	●		●						●		IBM, Cray
LAMPS	●		●								CDC, IBM
LASSAQ	●		●								DEC

N = Nonlinear analysis   [1] = Optimization   [2] = Torsion

range of applications. Also hardware information is included in the tables. A dot in the table means that the appropriate option is available in the program. Programs in a descriptive survey are ordered alphabetically.

Finite Element Software for Stress Analysis

TABLE 2

Program name	2D analysis	3D analysis	Macromech static	Macromech dynamic	Micromech static	Micromech dynamic	Heat transfer	Buckling	Fracture mechanics	Joints, bonding	Hardware
LAVIB	●			●N							CDC, DEC
LSHELL, NLSHELL		●	●N	●N							CDC
MULTPN	●				●						Any computer
NALCOM		●		●N							IBM, VAX, Apollo
NAPLES, VAPLES	●		●N	●				●		●	IBM
NOLIN	●		●N					●			IBM, Cray
NOLACS	●		●N		●N					●	IBM
PAC78		●	●N		●N				●	●	IBM, CDC, Univac
PLASCAN, PLASTOSH	●		●N								ICL, Cyber
Q3DG	●	●			●						Any computer
SANWAT	●		●								VAX
SAPCOM	●		●	●					●		IBM, DEC, Univac
SDSAP-CJ	●		●	●	●	●			●	●	IBM, HP
STAEBL [3]	●			●	●						IBM
SUN/FL	●	●	●N	●	●N		●	●	●		CDC, Cyber, VAX
THICK		●	●N						●		CDC, Cyber
THVROLL	●		●N								IBM, Amdahl
TSJ1 [2]				●						●	DEC
WANG	●	●							●		IBM, CDC, VAX
WAVE PROPAGATION	●			●							Hitachi

N = Nonlinear analysis   [3] = Blade design   [2] = Torsion

Program Name: *AB5U*

Program developer: M. C. Chen and L. N. O. Clewlow (preprocessor programs)

Source for program information: Aerojet Manufacturing Company, 601 South Placentia Ave., Fullerton, CA 92631, USA.

Program category: Program origin- industrial company. Status- fully operational. Application- macromechanical analysis, statics.

Program description: AB5U is a program for axisymmetric and plane structures having cylindrically anisotropic and nonlinear material properties. Displacement formulation, element library contains higher-order elements. The plastic analysis uses the successive approximation method and a bilinear stress-strain relationship. Program output: principal stresses and strains, effective stresses and strains for each element in both the composite and the liner. Preprocessors EPCS, FIBER and REVNOD calculates the geometric data and elastic constants necessary for input to AB5U. EPCS computes the composite elastic constants for each wrap and also for multilayered wraps. Composite thickness build-up at various angles along the meridian is calculated too. FIBER calculates the elastic constants for each element/ group of elements ready to proceed directly to the AB5U. REVNOD can quickly modify the composite winding pattern and node coordinates to facilitate design optimization. Postprocessor POST calculates the residual compressive stresses at zero pressure and the operating strain range. The burst failure prediction using maximum filament stress and maximum strain criteria is included in the program. Fatigue failure prediction.

Element library: Membrane, axisymmetric higher-order elements, slip elements used between the liner and composite to transfer only the normal load.

Hardware: Not known.

Program applications: Design of filament-reinforced metallic pressure vessels.

Program availability: AB5U is an "in-house" program.

Reference: MA-36

Program Name: APOC

Program developer: Yuan Yuan, P.O. Box 44, First Division, Anshun Guizhou Province, The People's Republic of China.

Source for program information: Tsun Kuei Wang, Beijing Institute of Aeronautics and Astronautics, Beijing, The People's Republic of China.

Program category: Program origin- university. Status- fully operational. Application- macromechanical anlaysis, interlayer analysis.

Program description: APOC is a program designed mainly to develop the elements for analysis of composite materials. Linear static analysis may be handled. The program is relatively small, written in FORTRAN IV. Program output: displacements and stresses in every layer, incl. the warp along the thickness and interlayer stresses.

Element library: Rectangular hybrid stress element; a group of rectangular hybrid stress elements specially developed for multilayer laminates. This element group, SZ-YS, appears to be a 2D like most elements used for plates, but it is formed in such a way that it can calculate the complex stress and displacement distributions like 3D elements.

Other program capabilities: Own routines can be inserted easily into the program, option for various output.

Hardware: Felix C-256 (made in Romania)

Program applications: Laminate design; investigation of influences of boundary conditions.

Program availability: Available for exchange.

Reference: EL-52, EL-55, MA-199, MA-208,209.

Finite Element Software for Stress Analysis   169

Program Name: *BEAM*

Program developer: A. R. Zak, Aero & Astro Engineering Dept., University of Illinois, Urbana, Illinois 61801, USA.

Source for program information: Same as for program developer.

Program category: Program origin- university. Status- fully operational. Application- macromechanical analysis, buckling.

Program description: BEAM is a program developed for the linear static analysis of beams. Open and closed beams can be handled. Thin, laminated cross-section is permitted. Composite beams of general configurations may be analysed. Beams may be subjected to mechanical or thermal loads. The program is incorporated in the general purpose finite element program NEPSAP.

Element library: Beam elements.

Other program capabilities: Option for various output, local coordinate systems.

Hardware: Cyber.

Program applications: Composite beams, general thin configurations.

Program availability: By special arrangements.

Program Name: *BEOS*

Program developer: K. Rohwer, DFVLR, Braunschweig, West Germany.

Source for program information: DFVLR, Inst. für Strukturmechanik, Flughafen D 3300 Braunschweig, West Germany.

Program category: Program origin- research institute. Status- fully operational. Application- macromechanical analysis, buckling.

Program description: BEOS (Buckling Loads of Eccentrically Orthotropic Sandwich Shells) is a program that calculates buckling loads, natural frequencies, and the initial postbuckling behaviour of layered, orthotropic and discretely stiffened shells. Stiffeners are permitted. The program calculates global buckling, not local buckling. Proportionality between load and deformation must be preserved. Shells are discretized into rectangular subregions. Program output: buckling load factor, buckling modes, natural frequencies and natural modes, initial postbuckling path.

Element library: Rectangular and parallelogram elements for plates and shells of Kirchhoff-Love or sandwich type.

Other program capabilities: Modelling is simple, standard boundary conditions may be generated; plot routines for buckling modes and natural modes (based on GINO-F); option for various output; free-field input.

Hardware: IBM, Univac, VAX 780.

Program applications: Airbus A310 vertical tail plane, solar panels.

Program availability: Can be purchased from the developer.

Program Name: *BOLT*

Program developer: Fu-Kuo Chang, Arizona State University, Dept. of Mechanical and Aerospace Engineering, Tempe, Arizona 85287, USA.

Source for program information: Same as for program developer.

Program category: Program origin- university. Status- fully operational. Application- fracture mechanics, bolted joints.

Program description: BOLT is a program for the analysis of bolted joints in laminated composites. The program can predict the strength and failure mode of multi-hole bolted joints. Chang-Scott-Springer failure criterion and Yamada-Sun failure criterion are included in the program. Program output: strength and failure mode of bolted joints.

Element library: Isoparametric quadrilateral elements.

Other program capabilities: Free-field input, restart.

Hardware: VAX, IBM, Amdahl.

Program applications: Bolted joint design.

Program availability: Available on request (nominal fee).

Reference: JB-23 to JB-27.

Program Name: CCMFE

Program developer: Center for Composite Materials, University of Delaware, 201 Spencer Laboratory, Newark, Delaware 19716, USA.

Source for program information: As above.

Program category: Program origin- university. Status- being developed. Application- micromechanical and macromechanical analysis, fracture mechanics, joints and bonding.

Program description: CCMFE is a program developed for the static linear and nonlinear analysis of composite materials. Steady-state heat transfer problems may also be handled. Nonlinear material models are available in conjunction with isotropic and orthotropic properties. Computation of material properties for long as well as for short fibers. The program will be included in the suite of CCM programs. CADOT is a finite element pre-/postprocessor for composite material design. It interfaces to commercial finite element codes SAP-5 and ADINA. Interface does not include all elements. CADOT offers Tresca, Tsai-Wu, and Generalized von Mises (Tsai-Hahn) failure criteria.

Element library: 2D plane stress/strain/axisymmetric elements with isotropic, orthotropic or shear-coupled material properties; 3D thin plate/shell element with both membrane and bending stiffness, fully anisotropic material properties include shear-coupling, bend-twist coupling and bending-extensional coupling; 3D solid element with shear-coupled material properties.

Other program capabilities: Pre-/postprocessors available, mesh generation, plotting.

Hardware: VAX, IBM.

Program applications: Program still in development.

Program availability: Only to sponsors of CCM (Center for Composite Materials).

Program Name: CLFE2D

Program developer: M. B. Buczek, M. A. Gregory, C. T. Herakovich.

Source for program information: Virginia Polytechnic Institute & State University, Engineering Science & Mechanics, att C. T. Herakovich, Blacksburg, VA 24061, USA.

Program category: Program origin- university. Status- fully operational. Application- macromechanical analysis.

Program description: CLFE2D is a 2D generalized plane strain program for laminated composites subjected to mechanical and hygrothermal loading. The program was developed to calculate the displacements, strains, stresses and strain energy

Finite Element Software for Stress Analysis 171

densities in a finite width composite laminate. It allows to input one set of 3D orthotropic material properties. The user can then specify the angle of material principal orientation for each element in the mesh. CLFM2D is a program for composite laminate fracture mechanics.

Element library: Linear, four node, general quadrilateral isoparametric element.

Other program capabilities: Undeformed and deformed finite element meshes may be plotted, option for various output.

Hardware: IBM.

Program applications: Analysis of laminated composites under mechanical and hygrothermal loading, fracture mechanics- crack growth characteristics.

Program availability: Available.

Reference: FM-28, FM-29, FM-30.

Program Name: *COMPASS*

Program developer: S. R. Lin, The Aerospace Corp, P.O. Box 92957, Los Angeles, California 90009, USA.

Source for program information: As above.

Program category: Program origin- industrial company. Status- being developed. Application- nonlinear static analysis.

Program description: COMPASS is developed for the elastic-plastic analysis of axisymmetric solid structures with unequal tension-compression and temperature dependent orthotropic material properties. Macromechanics plastic yielding criterion for anisotropic materials is included in the program. Generators for mesh, boundary conditions and loading are available. Mesh and stress contour plotting can be handled.

Element library: Axisymmetric ring, plane stress, plane strain elements.

Hardware: CDC.

Program applications: Program is still under development.

Program availability: Not for public release yet.

Program Name: *COMPOSITE*, Japan

Program developer: Y. Yamada, Institute of Industrial Science, University of Tokyo, 22-1, Roppongi 7-chome, Minato-ku, Tokyo, 106 Japan.

Source for program information: Y. Yamada or Nonlinear Analysis Program Research Assoc., Office at Japan Advanced Numerical Analysis Inc., 1-71, Nakamegro 1-chome, Megro-ku, Tokyo, 153 Japan.

Program category: Program origin- university. Status- fully operational. Application- nonlinear static analysis, buckling, instability, fracture mechanics.

Program description: COMPOSITE is a program for the inelastic small deformation static analysis. The program was originally developed for the analysis of composite materials but later it has been extended for the thermal-elastic-plastic-creep analysis of nuclear reactors/components. Generalized Voigt or Kelvin model is adopted. Isotropic and kinematic hardening and combination of both. The program allows for any type of stress rates that have appeared in the literature. Plastic buckling and/or instability problems can be handled.

Element library: 3D isoparametric element, crack-tip element.

Other program capabilities: Automatic renumbering of nodes, local coordinate

systems, contact problems can be handled, option for selective output.
Hardware: IBM, Hitac.
Program availability: Not known.
Reference: MA-224, MA-225, DY-47, EL-54.

Program Name: COMPOSITE/USA

Program developer: J. D. Lee, Dept. of Mechanical Engineering, University of Minnesota, Minneapolis, Minnesota 55455, USA.
Source for program information: As above.
Program category: Program origin- university. Status- fully operational. Application- macromechanical static analysis, fracture mechanics.
Program description: The program is capable of calculating the stress distribution, identifying the damage zone and mode of failure, analyzing the stress redistribution and the damage accumulation, and determining the ultimate strength of composite laminates. Linear anisotropic elastic continuum. A failure criterion which can identify the mode of failure-fiber breakage, matrix failure, delamination, is incorporated in the program. If the failure occurs, the stiffness matrix in the damage zone will be modified according to its mode of failure. Program output: stress distribution, damage zone, mode of failure, stress redistribution, progressive failure, ultimate strength of the laminate.
Element library: 8 node solid element.
Other program capabilities: Generator for boundary conditions and loading, local coordinate systems.
Hardware: STAR-100
Program applications: Analysis of fiber-reinforced composite laminate with a centered hole (NASA contract).
Program availability: Available, contact the program developer.
Reference: MA-107, FM-87.

Program Name: COSMOS

Program developer: Fuji Heavy Industries Ltd., Aircraft Engineering Div., 1-1-11 Yonan, Utsunomiya, Tochigi, 320 Japan.
Source for program information: As above.
Program category: Program origin. industrial company. Status- being developed Application- structural optimization.
Program description: COSMOS (Composite Structure Multi-Constraint Optimizing System) is a program for the optimization of composite structures. The program is able to size element dimension such as thickness and cross sectional area to minimize the structural weight under constraints of strength, deflection and flutter speed. Stress and displacements are analysed by the FEM module. Static and dynamic analysis may be performed. Isotropic and orthotropic material properties. Program output: stress and displacement of the optimum structure including element dimensions. The following criteria are included: maximum stress, maximum strain, Hoffman, Tsai-Wu. FEMAS is a preprocessor for generation of topology, boundary conditions and loading, not available for composite material properties.
Element library: Isoparametric plate, rod, pure shear panel.
Other program capabilities: Option for various output; plot routines available

for stress contour, stress vector, deformation, mode vector; own routines can be inserted.

Hardware: IBM.

Program applications: Jet trainer, passenger aircraft, research.

Program availability: In-house program, not for outside users.

Program Name: *EPT3*

Program developer: R. Y. Itani, Washington State University, Civil and Environmental Engineering, Pullman, Washington, USA.

Source for program information: Washington State University, Civil and Environmental Engineering, Pullman, Washington, USA.

Program category: Program origin- university. Status- fully operational. Application- torsion analysis of axisymmetric bars.

Program description: EPT3 is a program for the static analysis of elasto-plastic torsion of uniform and nonuniform nonhomogeneous bars. Each constituent material of a given bar is assumed to be isotropic and to behave in an elastic perfectly-plastic manner. The cross section of uniform bars may have any arbitrary shape including a hole. The nonuniform bars are restricted to be axisymmetric, they may also have an axisymmetric hole. No assumption or restrictions are made concerning plastic unloading. Analysed bars can be subjected to a quasi-static monotonically increasing torque. Problem solution is formulated as a quadratic programming through the use of FE approximation and application of a minimum rate principle of plasticity. Program limitation: 170 nodes, 200 elements.

Element library: Axisymmetric elements.

Other program capabilities: Mesh generator, plot routines are available.

Hardware: IBM.

Program applications: Torsion of nonuniform axisymmetric bars.

Program availability: Not known.

Reference: JB-47.

Program Name: *FE2D, FE3D*

Program developer: T. K. Kuppusamy, Civil Engineering Dept., VPI & State University, Blacksburg, VA 24061, USA.

Source for program information: As above.

Program category: Program origin- university. Status- fully operational. Application- micromechanical analysis, macromechanical analysis, linear and nonlinear static analysis, heat transfer.

Program description: Both 2D plane strain, axially symmetric and 3D versions are available with pre- and postprocessor capabilities. Linear and nonlinear static analysis as well as steady-state and transient heat transfer analysis may be handled. Material models available- nonlinear elastic, Ramberg-Osgood model, elastic-plastic. Elastic-plastic modified Hill's criterion is included in the program.

Element library: 8 to 21 node 3D brick elements, 4 to 8 node 2D quadrilateral elements.

Other program capabilities: Own routines can easily be inserted, plot routines are available.

Hardware: IBM.

Program applications: Analysis of composite plates.
Program availability: Not available for commercial use.
Reference: MA-96.

Program Name: *FIBER*

Program developer: L. J. Ebert, Metallurgy and Materials Science, Case Western Reserve University, Cleveland, Ohio. P. L. Flynn and P. K. Wright, General Electric Co., Evandale, Ohio, USA.

Source for program information: As above.

Program category: Program origin- university and industrial company. Status- fully operational. Application- micromechanical analysis.

Program description: FIBER is a special purpose program for the static, material nonlinear analysis. Tangent modulus description of nonlinear stress-strain response. FE model consists of basic symmetric element of fiber and matrix. Program output: stress, strain, strain energy. Specific options for composites: matrix yield and plastic strain, relation of macroscopic stress-strain behavior to local matrix stress-strain. The program is written in the Algol language.

Element library: Constant strain triangular element.

Other program capabilities: Free-field input, restart.

Hardware: Univac.

Program applications: Plastic flow prediction in notched fiber composite materials.

Program availability: Not known.

Reference: MI-19.

Program Name: *FRACOM*

Program developer: C. G. Aronsson and J. Bäcklund.

Source for program information: C. G. Aronsson, Dept. of Mechanical Engng. Linköping Institute of Technology, S-581 83 Linköping, Sweden. J. Bäcklund, Dept. of Aero Struct. & Matrls, The Royal Institute of Technology, S-100 44 Stockholm, Sweden.

Program category: Program origin- university. Status- fully operational. Application- fracture mechanics.

Program description: FRACOM calculates tensile fracture loads for notched laminates. Condensed stiffness matrix from FE program is required, together with unnotched tensile strength and fracture energy. Program output- displacements and extension of damage zone as functions of external load.

Element library: Depends on FE-program used.

Hardware: Nord, VAX, personal computers.

Program applications: JAS aircraft (for SAAB-SCANIA AB)

Program availability: To be available in 1986.

References: FM-12 to 17.

Program Name: *FRAMEAP*

Program developer: Wen-Hwa Chen, Dept. of Power Mechanical Engng., National Tsing Hua University, Hsinchu, Taiwan.

Source for program information: As above.

Program category: Program origin- university. Status- being developed. Application- fracture mechanics, joints and bonding, buckling, flutter and vibration.

Program description: FRAMEAP (Fracture Mechanics Analysis Program) is a program developed to deal with static and dynamic fracture problems. Several types of hybrid and conventional finite elements are included in the element library. Combination of different element types is possible. Evaluation of stress intensity factors and J-integrals. Static and dynamic response, steady-state and transient heat transfer problems, linear elastic and elasto-plastic, finite deformation fracture problems may be handled.

Element library: Bimaterial singular cracked elements for 2D and bending.

Other program capabilities: Own routines can easily be inserted, free-field input, local coordinate systems, option for various output. Plot routines are available.

Hardware: CDC, IBM, VAX.

Program applications: Analysis of fracture mechanics problems.

Program availability: Available.

Reference: FM-34 to 37.

Program Name: *GAMNAS*

Program developer: J. Whitcomb and B. Dattaguru.

Source for program information: J. Whitcomb, Mail Stop 188E, NASA Langley Research Center, Hampton, VA 23665, USA.

Program category: Program origin- research institute. Status- fully operational. Application- micromechanical analysis, fracture mechanics, joints and bonding.

Program description: The program performs linear and nonlinear static analysis of 2D structures. Geometric, material, and combined geometric and material nonlinearity is permitted. Plane stress or plane strain problems may be handled. Program output: displacements, stresses, strain-energy release rates.

Element library: 4 node quadrilateral elements with full and reduced integration.

Other program capabilities: Own routines can easily be inserted, option for various output.

Hardware: Prime.

Program applications: Analysis of bonded joints.

Program availability: Available through COSMIC.

Reference: JB-34, JB-37, JB-52.

Program Name: *LAMINA*

Program developer: H. Eggers, Inst. für Strukturmechanik, DFVLA Braunschweig, Flughafen, 3300 Braunschweig, West Germany.

Source for program information: As above.

Program category: Program origin- research institute. Status- being developed. Application- macromechanical analysis, fracture mechanics.

Program description: LAMINA is a program for the macromechanical analysis of laminate composites. Layered elements for laminates are built up by the substructure technique. Single elements include unidirectional fiber orientations only. The program is primarily designed to calculate physical nonlinear problems which will be introduced in the future. Arbitrary coupling of the DOF's of the elements, nonpositive algebraic systems, optimization of the solution process. Program output- complete solution vector, displacements and stresses in the nodes of a layered element, displacements and stresses transformed to a local coordinate system at a given arbitrary located point in the structure, J integrals according to Rice.

Element library: A rectangular element with 9 nodes for the plane stress state based on the Hellinger/Reissner's principle.

Other program capabilities: Local coordinate systems. Substructuring technique. Option for various output. Node renumbering.

Hardware: IBM, Cray.

Program applications: ESA-project, calculation of fracture maps for the carbon fiber reinforced plastics.

Program availability: First release in 1986.

Program Name: LAMPS

Program developer: Dept. of the NAVY, NSRDC, Carderrock Labs., Maryland 20034, USA.

Source for program information: The same as for program developer.

Program category: Program origin- research institute. Status- fully operational. Application- linear static analysis of plates and shells.

Program description: LAMPS program is developed for the static analysis of laminated orthotropic plates and shallow shells. Linear elastic material only. Homogeneous elements are extended to laminated elements where each lamina may have different thickness and material properties. Program limitation: 64 elements, 85 nodes and 10 laminas.

Element library: Plate and shell elements.

Hardware: CDC, IBM.

Program applications: Analysis of laminated orthotropic plates and shells.

Program availability: Not known.

Program Name: LASSAQ

Program developer: P. V. Ramanamurthy, K. P. Rao and A. Venkatesh.

Source for program information: Centre for Scientific and Industrial Consultancy, Indian Institute of Science, Bangalore-12, India (att. K. P. Rao).

Program category: Program origin- university. Status- fully operational. Application- stress analysis of stiffened laminated anisotropic shells.

Program description: LASSAQ is a 2D code for the stress analysis of stiffened laminated anisotropic shells. The program uses the doubly curved quadrilateral laminated anisotropic shell of revolution and its compatible meridional and parallel circle stiffener elements. Shell thickness or stiffener may be made up of an arbitrary number of bonded layers, each of which is linear or isotropic (unimodulus or bimodulus) with an arbitrary orientation of principal axes. Program output- displacements at nodes, strains at all Gauss points.

Element library: Quadrilateral laminated anisotropic thin shell of revolution. Laminated anisotropic meridional stiffener element. Laminated anisotropic parallel circle stiffener element.

Hardware: DEC 1090.

Program applications: Analysis of stiffened laminated anisotropic shells.

Program availability: Available. Contact the program developer.

Reference: MA-158, MA-201, EL-46 to 50.

Program Name: <u>LAVIB</u>

Program developer: Chuh Mei, Dept. of Mechanical Engng. and Mechanics, Old Dominion University, Norfolk, VA 23508, USA.

Source for program information: The same as for program developer.

Program category: Program origin- university. Status- fully operational for free and forced nonlinear vibrations, under development for nonlinear random vibrations. Application- large-amplitude vibrations of beams and plates.

Program description: LAVIB is a special purpose 2D program developed for large-amplitude vibration analysis of symmetrically laminated composite beams and plates. Free and forced nonlinear vibrations may be handled. Amplitude-frequency relation can be determined. Program output: nonlinear frequency, strain.

Element library: 24 DOF rectangular plate element, 18 DOF triangular plate element.

Other program capabilities: Own routines can easily be inserted into the program.

Hardware: CDC, DEC.

Program applications: Large-amplitude vibrations of beams and plates.

Program availability: Not known.

Reference: DY-18, DY-19.

Program Name: <u>LSHELL and NLSHELL</u>

Program developer: H. P. Huttelmaier and M. Epstein.

Source for program information: Dept. of Mechanical Engineering, University of Calgary, Alberta, Canada T2N 1N4.

Program category: Program origin- university. Status- being developed. Application- linear and nonlinear analysis of multilayered shells.

Program description: LSHELL is a program for 3D linear analysis of multilayered shells. NLSHELL is a program for 3D nonlinear analysis of multilayered shells. Static and dynamic problems can be handled. Geometrically nonlinear analysis only. Lagrangian based nonlinear formulation of a multilayered shell. Theory used allows for piecewise linear displacement fields through the thickness. Applicable to very thick shells. General anisotropic linear elastic material properties. May be varied from layer to layer. Program output- nodal displacements, director displacements, stresses at nodal points and at element center (at center of each layer and at interfaces).

Element library: Multilayered element, bilinear interpolation functions, under integration of shear terms.

Other program capabilities: Restart option.

Hardware: CDC.

Program applications: Analysis of multilayered plates and shells.

Program availability: Program available at current status at no charge.

Reference: EL-17, MA-57, MA-58

Program Name: MULTPN

Program developer: J. A. Mandel, Dept. of Civil Engineering, Syracuse University, Syracuse, N Y 13210, USA.

Source for program information: The same as for program developer.

Program category: Program origin- university. Status- fully operational. Application- micromechanical analysis, fracture mechanics.

Program description: MULTPN is a 2D code for the multiphase micromechanical analysis of fiber reinforced composites. Static linear elastic problems may be handled. Concentrated and distributed mechanical loads are permitted. Program output- displacements, stresses, stress intensity factors.

Element library: 2D quadratic elements, singular quarter-point elements, quadratic interface element, singular quadratic interface element, 16 node connector element for multiplane analysis.

Other program capabilities: Load generators.

Hardware: Any computer with FORTRAN IV compiler.

Program applications: Crack growth study in composites.

Program availability: Information available upon request.

Reference: MI-40.

Program Name: NALCOM

Program developer: O. H. Griffin, Jr, 6142 Waiting Spring, Columbia, MD 21045, USA.

Source for program information: The same as for program developer.

Program category: Program origin- industrial company. Status- fully operational. Application- linear and nonlinear analysis of laminated composites, micromechanical analysis.

Program description: NALCOM is a program for the static linear and nonlinear analysis of laminated composites. Material properties may be isotropic or transversely isotropic. All material properties may vary with temperature. Both first and second order nonlinear temperature effects are handled. The material may undergo elastic or elastic-plastic behaviour. The plasticity model uses a modified Hill-type yield criterion and flow rule. Nonlinear stress-strain data is input as Ramberg-Osgood form. Loads may be thermal, concentrated forces and specified displacements.

Element library: 24 node, 72 DOF isoparametric solid element.

Other program capabilities: Own routines can easily be inserted into the program. Option for various output. Plot routines are available.

Hardware: IBM, VAX, Apollo.

Program applications: Study of interlaminar stress distributions in laminates.

Program availability: Contact developer.

Reference: MA-67 to 69.

Finite Element Software for Stress Analysis 179

Program Name: *NAPLES* and *VAPLES*

Program developer: J. N. Reddy, 505 Cranwell Circle, Blacksburg, VA 24060, USA.

Source for program information: J. N. Reddy, ESM Dept., Room 220, Norris Hall, Virginia Polytechnic Institute, Blacksburg, VA 24061, USA.

Program category: Program origin- university. Status- fully operational. Application- nonlinear static, dynamic and vibration analysis, buckling, impact, joints and bonding.

Program description: NAPLES is the nonlinear analysis program for laminated elastic structures. VAPLES is the program for vibration analysis of laminated elastic structures. Linear and geometrically nonlinear analyses may be handled. Arbitrary number of layers, material options, and stacking sequences is permitted. NAPLES is based on a shear deformation theory of laminated shells (Sanders' version) and accounts for geometric nonlinearity (in the von Karman sense) and transient response. Program output- displacements and stresses for each load and/or time step in each lamina. The program can be inserted as a module into another general purpose FE program that has pre- and postprocessors, etc.

Element library: Linear and quadratic rectangular isoparametric elements. Plate and doubly-curved shell elements.

Other program capabilities: Mesh generator for a rectangular geometry. Own routines can be accommodated into the program.

Hardware: IBM.

Program applications: Analysis of laminated plates, shells and beams.

Program availability: The program is for sale, price is negotiable.

Reference: MA-162 to 170, DY-30 to 37, EL-38.

Program Name: *NOLIN*

Program developer: K. Rohwer, DFVLR, Inst. für Strukturmechanik, Flughafen, D-3300 Braunschweig, West Germany.

Source for program information: DFVLR, Inst. für Strukturmechanik, Flughafen, D-3300 Braunschweig, West Germany.

Program category: Program origin- research institute. Status- being developed. Application- static nonlinear analysis, buckling.

Program description: NOLIN is a program for the geometric nonlinear analysis of layered, anisotropic curved panels. Their behaviours can be described by up to five functional DOF. Quadratic stress-strain relations. The complete stiffness matrix including all coupling coefficients may be inputted.

Element library: Rectangular and parallelogram elements for plates and shells.

Other program capabilities: Boundary conditions and loads can be generated. Free-field input. Option for various output. Restart capability. Plot routines are available.

Hardware: IBM, Cray.

Program applications: Only research applications yet.

Program availability: The program is under development.

Reference: MA-174, MA-175.

Program Name: *NOLACS*

Program developer: C. T. Herakovich *et al.*

Source for program information: C. T. Herakovich, Dept. of Engng. Science and Mechanics, Virginia Polytechnic Institute and State University, Blacksburg, VA 24601, USA.

Program category: Program origin- university. Status- fully operational. Application- linear and nonlinear static analysis, joints and bonding.

Program description: NOLACS is a program for the linear and nonlinear static analysis of laminated composites. The program is an improved version of NONCOM and NONCOM1. It utilizes the full 3D nonlinear material properties, inclusive hygrothermal effects. A modular form of NOLACS uses a low-speed storage to greatly reduce the problem size restrictions of its predecessors. Interlaminar stresses can be determined for mechanical loading, uniform hygrothermal loading, and gradient moisture loading. Nonlinear procedure is based on the incremental procedure. Determination of tangent modulus with Ramberg-Osgood approximations. The gradients representing actual moisture distributions can be generated by a second order, 2D finite difference program HYDIP.

Element library: Constant strain triangular elements.

Hardware: IBM.

Program applications: Study of interlaminar stresses near free edges.

Program availability: Not known.

Reference: MI-18, MA-171, JB-46.

Program Name: PAC78

Program developer: Yehia A Bahei-El-Din, Structural Engineering Dept., Cairo University, Giza, Egypt.

Source for program information: As above.

Program category: Program origin- university. Status- fully operational. Application- macromechanical and micromechanical analysis, damage, joints and bonding.

Program description: PAC78 is a program for the linear and nonlinear static analysis. Material nonlinearities only. The nonlinearities caused by the elastic-plastic behaviour of the composites are handled by a modified Newton-Raphson iteration procedure. Laminated structures with layers in several directions may be handled. Each element is treated as macroscopically homogeneous and anisotropic. The matrix may be nonhardening or kinematic hardening. The von Mises yield criterion, Prager's hardening rule with Ziegler's modification. Fiber failure criterion has been added to the program. This version is named as PAFAC and is available through the COSMIC. Static mechanical loads with any loading, unloading/reloading sequence. Program output- local and global stresses and strains, forces at constrained nodes at each stage of loading and unloading.

Element library: 8 node hexahedron solid element. By means of deflection boundary conditions one may simulate beams, plates, plane strain elements, etc.

Other program capabilities: User may provide own subroutines for mesh generation and generation of boundary conditions. A restart capability is available. Option for various output. Local coordinate systems. Own routines can easily be inserted into the program.

Hardware: IBM, CDC, Univac.

Program applications: Stress analysis of laminated plates containing holes and notches, including prediction of initial yielding and initial fiber failure.

Program availability: PAC78 is available through membership of PAC78 Users Group

Finite Element Software for Stress Analysis   181

(S Utku, Dept. of Civil Engng., Duke University, Durham NC 27706, USA).
PAFAC is available through the COSMIC.

Reference: MA-10 to 16.

Program Name: PLASCAN and PLASTOSHELL

Program developer: J. A. Figueiras and D. R. J. Owen.

Source for program information: J. A. Figueiras, Faculdade de Engenharia, Rua dos Bragas, 4099 Porto, Portugal. D. R. J. Owen, Civil Engng. Dept., University College of Swansea, Swansea, UK.

Program category: Program origin- university. Status- fully operational. Application- linear and nonlinear static analysis of shell structures.

Program description: PLASCAN is a program for the elasto-plastic analysis of shell structures of fiber composite material with anisotropic material properties using the semiloof shell element. PLASTOSHELL is a program for the elastic-plastic analysis of thin and thick shell structures of laminate composite material using quadratic thick shell elements. Layered formulation. Static analysis, material and geometrical nonlinearities are permitted. Anisotropic yield criterion with anisotropic work-hardening parameters is incorporated. The formulation is based on the flow theory of plasticity. Program output- displacements, reactions, stresses at each Gauss point and at each layer through the thickness, stress resultants at each Gauss point.

Element library: Semiloof shell element. 8 node Serend., heterosis, 9 node Lagrangian thick shell elements.

Other program capabilities: Own routines can be inserted. Option for various output. Local coordinate system. Restart.

Hardware: ICL, Cyber.

Program applications: Analysis of fiber reinforced and laminated composite plates and shells. Stress analysis of a wind turbine blade.

Program availability: Available under conditions specified for each specific case.

Reference: MA-60, MA-137, MA-138.

Program Name: Q3DG

Program developer: I. S. Raju, Mail Stop 188E, NASA Langley Research Center, Hampton, VA 23665, USA.

Source for program information: I. S. Raju, Analytical Services and Materials, Inc., 103 Winder Road, Tabb, VA 23602, USA.

Program category: Program origin- research institute. Status- fully operational. Application- micromechanical analysis.

Program description: Q3DG computes the strain-energy release rates for delaminated composite laminates under mechanical, thermal and hygroscopic loadings. Strain-energy release rates are computed by virtual crack extension method. Fiber reinforced composites may be analysed. Ply-by-ply idealisation is used. Each ply of the composite is idealised as an equivalent-orthotropic material. Program output- displacements, stresses at each mode in the model and the strain-energy release rates (Mode I, Mode II, and Mode III) for delamination growth.

Element library: 8 node isoparametric quadrilateral element.

Other program capabilities: Rectangular mesh generator is available. Own routines can easily be inserted into the program. Option for various output.

Hardware: Most computers which use FORTRAN IV.

Program applications: Analysis of interlaminar stresses. 2D and 3D analyses of composite joints. Free-edge stress analysis.

Program availability: Available through the COSMIC.

Reference: MI-44, MI-46, MI-54, MI-55, JB-61, MA-216, MA-217.

Program Name: *SANWAT*

Program developer: R. Chandhuri and P. Seide.

Source for program information: R. Chandhuri, Dept. of Civil Engineering, University of Utah, Salt Lake City, Utah, USA. P. Seide, Dept. of Civil Engineering, University of Southern California, Los Angeles, CA 90089-1114, USA.

Program category: Program origin- university. Status- being developed. Application- linear static analysis of reinforced plates and shells.

Program description: SANWAT is a program for the analysis of reinforced laminated plates and shells. Linear static analysis only. The program uses quadratic triangular elements for each layer which are stacked. Shear deformations in each layer are permitted. Tangential displacements at interface of layers are the unknowns. Transverse displacement is constant through the thickness unless parts of layers are missing. Arbitrary orthotropic orientation and properties in each layer are permitted.

Element library: Plate, circular cylinder and sphere triangular elements.

Other program capabilities: None.

Hardware: VAX.

Program applications: Only applied to specific academic problems in dissertation including perforated plates and shells with partial and complete thickness holes.

Program availability: Not ready yet. Program listing is available.

Reference: MA-31, MA-185.

Program Name: *SAPCOM*

Program developer: Scientists of Structural Division, National Aeronautical Laboratory, Bangalore 560017, India.

Source for program information: Structural Division, National Aeronautical Laboratory, Bangalore 560017, India.

Program category: Program origin- research institute. Status- being developed. Application- linear static and dynamic analysis, fracture mechanics.

Program description: SAPCOM is a collection of special purpose programs for the finite element analysis of laminated composite plates and shells. The programs are developed to support the research work in the field of composite structural mechanics. Linear static and dynamic problems may be handled. General laminates taking into account material anisotropy, bending extensional coupling, transverse shear deformation effects and thermal effects may be analysed. Program output- nodal displacements, element stress resultants/layer, stresses, buckling loads, natural frequencies.

Element library: A family of triangular and quadrilateral elements for stretching and bending of plates, shells of revolution and shallow shells.

Other program capabilities: Own routines can be incorporated.

Hardware: IBM, DEC, Univac.

Program applications: Programs are used to support design of composite structures for aerospace applications, to study the behaviour of structural elements, etc.

Program availability: Not known.

Reference: MA-99 to 104, DY-14, EL-20, EL-21, FM-82 to 85, JB-49.

Program Name: *SDSAP-CJ*

Program developer: Toru Fujii, Faculty of Engineering, Doshisha University, Karasuma Imadegawa, Kamigyo-ku, Kyoto, Japan.

Source for program information: S. Amijima, Faculty of Engineering, Doshisha University, Karasuma Imadegawa, Kamigyo-ku, Kyoto, Japan.

Program category: Program origin- university. Status- fully operational. Application- micromechanical analysis, joints and bonding, impact.

Program description: SDSAP contains several programs which can be operational individually. A few of them are under development. Linear and nonlinear static and dynamic problems can be handled. Material library includes linear elastic isotropic and orthotropic models, and nonlinear isotropic adhesives. Some programs are written in FORTRAN, others in HP BASIC.

Element library: Beam element, plane stress membrane element, plate.

Other program capabilities: Own routines can be inserted.

Hardware: IBM, HP.

Program applications: Interlaminar dynamic stress analysis for two layered composites subjected to impact loading. Adhesive bonded stress analysis for composite joints subjected to in-plane and out-of-plane loads.

Program availability: Source lists are available under no options.

Reference: DY-2, JB-5 to 10.

Program Name: *STAEBL*

Program developer: K. Brown, Pratt & Whitney Aircraft, MS 163-10, East Hartford, Conn., USA.

Source for program information: The same as for the program developer.

Program category: Program origin- industrial company. Status- fully operational. Application- blade design, aerospace industry.

Program description: STAEBL is a special purpose program for blade design. Static problems, static problems with pre-stress and free vibration analysis may be handled. In-core solution for computer efficiency. Differential stiffness and centrifugal load option is included. Tsai-Wu equivalent stress criterion.

Element library: Triangular plate element with composite capability.

Other program capabilities: Local coordinate systems.

Hardware: IBM.

Program applications: Optimization of blade design.

Program availability: Available through NASA.

Reference: MA-23.

Program Name: *SUN/FL*

Program developer: C.T. Sun and D. H. Lin.

Source for program information: Dept. of Engineering Science, 231 Aero, University of Florida, Gainesville, Florida 32611, USA.

Program category: Program origin- university. Status- being developed. Application- micromechanical and macromechanical static analysis, buckling, impact, damping analysis, fracture mechanics.

Program description: The program can calculate 3D interlaminar stresses in composite plates. Any laminate design may be handled. Hybrid formulation is used. The method is characterized by an assumed stress field in the element and by an assumed displacement field along the interelement boundary. Static and dynamic problems can be analysed. Large deformations are permitted.

Element library: 4 node and 9 node isoparametric plate bending and plane element, 20 node solid element.

Other program capabilities: Free-field input. Local coordinate systems. Substructuring is available. Option for various output.

Hardware: CDC, Cyber, VAX.

Program applications: Free edge stress analysis, impact rupture, dynamic crack propagation, post-buckling of laminate plates, transient thermal stress analysis.

Program availability: The program is under development.

Reference: DY-40, FM-127.

Program Name: THICK

Program developer: A. Palazotto and R. Henrichsen.

Source for program information: Wright-Patterson AFB, AFIT/ENY, Ohio 45433, USA.

Program category: Program origin- research institute. Status- being developed. Application- buckling analysis.

Program description: THICK is a special purpose 3D program for the analysis of thick composite plates and shells. Static linear and nonlinear problems may be handled. The program will be incorporated into the general purpose program STAGSCI.

Element library: Triangular and quadrilateral plate elements.

Other program capabilities: Free-field input.

Hardware: CDC, Cyber.

Program applications: Analysis of thick composite plates and shells.

Program availability: The program is under development.

Reference: FM-47, BU-8, MA-220.

Program Name: THVROLL

Program developer: R. C. Batra, Dept. of Engineering Mechanics, University of Missouri-Rolla, Rolla, MO 65401-0249, USA.

Source for program information: The same as for the program developer.

Program category: Program origin- university. Status- fully operational. Application- nonlinear thermoviscoelastic analysis.

Program description: THVROLL is a program that analyses plane strain large deformations of composite materials made of a Boltzmann type viscoelastic material and a nonlinear elastic material. Both distributed pressure and shear tractions can be prescribed. Also point loads can be inputted.

Program output- displacements at nodal points, pressure field within each element, stresses and strains at each Gauss point, principal stresses and principal strains at Gauss points.

Element library: 4 node isoparametric quadrilateral elements.

Other program capabilities: Option for various output.

Hardware: IBM, Amdahl.

Program applications: Cold rolling of a plywood sheet, indentation of uniformly rotating roll covers.

Program availability: Program available for lease and/or sale.

Reference: MA-21.

Program Name: <u>TSJ1</u>

Program developer: E. J. Coyle, EI Du Pont Co., Wilmington, DE 19898, USA.

Source for program information: R. Byron Pipes, University of Delaware, Newark, Delaware, USA.

Program category: Program origin- university. Status- fully operational. Application- torsion analysis of shafts, joints and bonding.

Program description: The program is designed for the evaluation of axisymmetric and inhomogeneous orthotropic shafts of variable cross sections subjected to torsional loads.

Element library: 6 node triangular elements with quadratic shape functions.

Other program capabilities: Mesh generator is available. Own routines can easily be inserted. Option for various output. Plot routines for mesh and stress contours.

Hardware: DEC.

Program applications: Analysis of automotive drive shafts.

Program availability: Available.

Reference: JB-33.

Program Name: <u>WANG</u>

Program developer: A. S. D. Wang and students.

Source for program information: A. S. D. Wang, Drexel University, Philadelphia, PA 19104, USA.

Program category: Program origin- university. Status- fully operational. Application- static linear fracture mechanics analysis.

Program description: The program was developed for the simulation of plane crack growth under statically applied loads. Crack growth contour can be two-dimensional. All types of unidirectional fiber reinforced polymer systems and their laminates may be handled. Failure criteria for crack growth are included and can also be changed. Program output- stress, strain, strain energy density, strain energy release rate, stress distribution contours, etc.

Element library: 2D triangular and quadrilateral. 8 node and 21 node 3D solid element.

Other program capabilities: Generator for moving boundaries and loading changes. Plot routines for all output quantities. Free-field input. Substructuring technique is incorporated. Option for various output. Restart capability.

Hardware: IBM, CDC, VAX.

Program applications: Transverse crack growth, delamination growth and their interacting growth effects.

Program availability: To be used at Drexel University.

Reference: FM-40 to 42, FM-132 to 139.

Program Name: *WAVE PROPAGATION*

Program developer: S. Minagawa and M. Yamada, Denkitsushin University, Chofu, Tokyo 182, Japan.

Source for program information: The same as for the program developer.

Program category: Program origin- university. Status- fully operational. Application- propagation of waves in composites.

Program description: The program was developed to analyse the harmonic wave propagation in layered and fibre-reinforced composites. Piecewise linear approximating functions are used for the displacement and stress fields in a mixed variational formulation proposed in the form of a new quotient. Static, steady-state analyssis may be performed.

Element library: Triangular elements.

Other program capabilities: Generator for quasi-periodicity condition is available.

Hardware: Hitachi M-170.

Program applications: Analysis of harmonic waves in layered and fibre-reinforced composites.

Program availability: Not known.

Reference: DY-20.

ACKNOWLEDGEMENT

It should be emphasized that the list of presented programs is far from complete and it is possible that some capabilities of described programs are omitted. The author would appreciate to receive any comments on the paper and information about programs not included in this paper.

Sincere thanks are expressed to all contributors answering the questionnaires.

REFERENCES

1. Brebbia, C. A. ed., *Finite Element Systems, A Handbook,* Springer-Verlag, Berlin, 1982.

2. Fredriksson, B. and Mackerle, J. *Structural Mechanics Finite Element Computer Programs,* 4th Edition, AEC Co., Box 3044, Linköping, Sweden, 1983.

3. Perronne, N. and Pilkey, W. *Structural Mechanics Software Series,* vol. *1-5* University Press of Virginia, Charlottesville, VA.

4. Griffin, O. H., Jr. Evaluation of Finite-Element Software Packages for Stress Analysis of Laminated Composites. *Composite Technology Review,* Winter 1982, pp. 136-141.

APPENDIX

References

The list of references on the application of the finite element method for stress analysis of composite materials/structures is for a "quick" dissemination of relevant information divided into the following categories:

- macromechanical analysis (MA)
- micromechanical analysis (MI)
- buckling, post-buckling (BU)
- dynamic analysis (DY)
- element library (EL)
- fracture mechanics analysis (FM)
- joints, bonding (JB)

References listed are within each category arranged in an author alphabetical order. They have been extracted from the finite element structural mechanics data base MAKEBASE being currently developed by the author at the Linköping Institute of Technology.

Macromechanical Analysis (MA)

MA- 1 Adams D. F.,
  Inelastic Analysis of a Unidirectional Composite Subjected to Transverse Normal Loading, *J. Comp. Mat.*, Vol. 4, 1970, 310-328
MA- 2 Adams D. E.,
  Influence of the Polymer Matrix on the Mechanical Response of a Unidirectional Composite, 4th Int. Conf. Comp. Mat., Tokyo, 1982, 507-517
MA- 3 Agarwal B. D. and Bansal R. K.,
  Effect of an Interfacial Layer on the Properties of Fibrous Composites: A Theoretical Analysis, *Fibre Sci. Tech.* Vol. 12, No. 2, 1979, 149-158
MA- 4 Agarwal B. D. and et al.,
  Stress Distribution Around Hole of Angle-Ply Laminates under Uni-Axial Tension, 4th Int. Conf. Comp. Mat., Tokyo, 1982, 397-404
MA- 6 Akbarzadeh A.,
  Stress Analysis of Composite Materials by the Hybrid Finite Element Method, Ph D Thesis, Univ. of Wyoming, 1975
MA- 7 Al-Bazzaz A. J.,
  The Use of F.E.M. in Approximate Methods for the Analysis of Composite Double-Layer Space-Grid Structures, 4th Int. Conf. Austral. FEM, Melbourne, Aug. 1982, 149-152
MA- 8 Backlund J.,
  A Finite Element Analysis of Timber Beams Reinforced with Composite Material, Linkoping Inst. of Tech., IKP-R-139, Sweden, Aug. 1979
MA- 9 Backlund J. and Mackerle J.,
  Stress Analysis of Composite-Material Plates by Isoparametric Finite Element, Linkoping Inst. of Techn., IKP-R-130, Sweden, March, 1979
MA- 10 Bahei-El-din Y. A. and Dvorak G. J.,
  Plasticity Analysis of Laminated Composite Plates, in *Adv. Aerosp. Str. Mater.* ASME, 1982, 139-140
MA- 11 Bahei-El-Din Y. A. and Dvorak G. J.
  Plasticity Analysis of Laminated Composite Plates,
  J. Appl. Mechanics, ASME, Vol. 49, 1982, 740-746
MA- 12 Bahei-El-Din Y. A. and et al.,
  Finite Element Analysis of Elastic-Plastic Fibrous Composite Structures, Symp. Comp. Meth. Nonlin. Str. Solid Mech., Washington, D.C., 1980

MA- 13 Bahei-El-Din Y. A.,
Plastic Analysis of Metal-Matrix-Composite Laminates, Ph D Dissert., Duke University, 1979
MA- 14 Bahei-el-Din Y. A. and Dvorak G. J.,
Plastic Yielding at a Circular Hole in a Laminated FP-Al Plate, in *Modern Dev. Comp. Mat. Struct.*, ed. J. Vinson, 1979, 123-147
MA- 15 Bahei-el-Din Y. A. and Dvorak G. J.,
Plastic Deformation of a Laminated Plate with a Hole, J. Appl. Mech. ASME, Vol. *47*, 1980, 827-832
MA- 16 Bahei-el-Din Y. A. and et al.,
Finite Element Analysis of Elastic-Plastic Fibrous Composite Structures, *Computers & Structures*, Vol. *13*, 1981, 321-330
MA- 17 Baier H. J. and Schneermann M. W.,
Analysis and Synthesis of High Precision Structures, Aachen, DGLR-81-053, 1981
MA- 18 Barker R. M. and et al.,
Stress Concentrations Near Holes in Laminates, Proc. ASCE, Vol. *100*, EM3, 1974, 477-488
MA- 19 Barker R. M. and et al.,
Three-Dimensional Finite Element Analysis of Laminated Composites, Computers & Structures, Vol. *2*, 1972, 1013-29
MA- 20 Barker R. M. and et al.,
Three Dimensional Analysis of Stress Concentrations Near Holes in Laminated Composites, Conf. ASCE Spec. Comp. Mat., Pittsburgh, 1972
MA- 21 Batra R. C.,
Finite Deformations of a Composite Viscoelastic Roll Cover, 12th Southeast Conf. Theor. Appl. Mech., Aubum Un., 1984
MA-22 Boland P. L.,
Finite Element Solution for Anisotropic Plates, S M Thesis, MIT, Dept. of Aeron., 1971
MA-23 Brown K. W. and et al.,
Structural Tailoring of Engine Blades (STAEBL), 24th Str., Str. Dyn. Mater. Conf., Lake Tahoe, Nevada, May 1983
MA- 24 Bubeck R. B.,
Transient Thermal Stress Analysis of Composite Structures Including Continuously Varying Properties, Master Thesis, Naval Postgr. School, Monterey, 1975
MA- 25 Chamis C. C. and Lark R. F.,
Hybrid Composites - State-of-the-Art Review: Analysis, Design, Application and Fabrication, NASA Rep. TM X- 73545, 1977
MA- 26 Chamis C. C. and Minich M. D.,
Structural Response of a Fiber Composite Compressor Fan Blade Airfoil, NASA TM X-71623, 1975
MA- 27 Chamis C. C. and Sinclair J. H.,
Mechanics of Intraply Composites - Properties, Analysis and Design, Polymer Comp., Vol. *1*, 1980, 7-13
MA- 28 Chamis C. C. and Sinclair J. H.,
Mechanical Behaviour and Fracture Characteristics of Off-Axis Fiber Composites, Part II - Theory and Comparisons, NASA Rep. TP-1082, 1978
MA- 29 Chamis C. C. and Sinclair J. H.,
Ten-Degree Off-Axis Test for Shear Properties in Fiber Composites, *Experimental Mech.*, Vol. *17*, No. 9, 1977, 339
MA- 30 Chamis C. C. and et al.,
NASTRAN as an Analytical Research Tool for Composite Mechanics and Composite Structures, NASTRAN Users' Exp., NASA TM X-3428, Oct. 1976, 381-417
MA- 31 Chandhuri R.,
Static Analysis of Fiber Reinforced Laminated Plates and Shells with Shear Deformation Using Quadratic Triangular Elements, Dissert, Dept. of Civ. Eng., Un South Calif., L. Angeles, May 1973
MA- 32 Chang F. K. and et al.,
Design of Composite Laminates Containing Pin Loaded Holes, *J. Comp. Mater.*, May 1984, 279-289
MA- 33 Chang F. K. and et al.,

The Effect of Laminate Confjguration on Characteristic Lengths and Rail Shear Strength, *J. Comp. Mater.*, May 1984, 290-296

MA- 34 Chang T. Y. and Sawamiphakdi K.,
Large Deformation Analysis of Laminated Shells by Finite Element Method, Computers & Structures, Vol. *13*, 1981, 331-340

MA- 35 Chaudhary V. P.,
Finite Element Analysis of Axisymmetrical Shells of Composite Material, Ph D Thesis, Univ. of Nebraska, Lincoln, 1970

MA- 36 Chen M. C. and Clewlow L. N. O.,
Computer Analysis of Filament-Reinforced Metallic-Spherical Pressure Vessels, *Computers & Structures*, Vol. *7*, 1977, 93-102

MA- 37 Chung T. J.,
Thermomechanical Response of Inelastic Fibre Composites, *Int. J. Num. Meth. Engng.*, Vol. *9*, No. 1, 1975, 169-185

MA- 38 Chung T. J. and Eidson R. I.,
Static and Dynamic Analysis of Viscoelastic Fiber-Reinforced Composite Shells in Missile Structures, Alabama Univ., Dept. of Mechanical Eng., 1973

MA- 39 Chung T. J. and Prater J. L.,
A Constitutive Theory for Anisotropic Hygrothermoelasticity with Finite Element Applications, *J. Thermal Str.*, Vol. *3*, 1980, 435-452

MA- 40 Crane D. A. and Adams D. F.,
Finite Element Analysis of a Unidirectional Composite Including Longitudinal Shear Loading, Univ. of Wyoming, UWME-DR-001-104-1, Dec. 1980

MA- 41 Crossmann F. W.,
Computer Simulation of the Deformation of Composite Materials by Finite Element Analysis, Ph D Thesis, Stanford Univ., 1972

MA- 42 Crossman F. W. and Karlak R. E.,
Creep of B/Al Composites as Influenced by Residual Stresses, Bond Sgrength and Fiber Packing Geometry, 6th AIME Annual Meet., Univ. of Pittsburg, 1974

MA- 43 Dal P. H. and Wouters C. H.,
Etude du Comportement Non Lineaire de Structures Spatiales Composites par la Methode des Elements Finis, Univ. Libre de Bruxelles, 1979

MA- 44 Dana J. R.,
Three Dimensional Finite Element Analysis of Thick Laminated Composites Including Interlaminar and Boundary Effects Near Circular Holes, Ph D Diss, Virginia Pol. Inst. State Un., Blacksburg, 1973

MA- 45 Dana J. R. and Barker R. M.,
Three Dimensional Analysis for the Stress Distribution Near Circular Holes in Laminated Composites, Virginia Polyt. Inst. State Univ., VPI-E-74-18, 1974

MA- 46 Davis R. C.
Stress Analysis and Buckling of J-Stiffened Graphite-Epoxy Panel, NASA Rep. TP-1607, 1970

MA- 47 Delcourt-Bon C. and Laschet G. Essais de caracterisation des materiaux composites 3-D, LTAS Rep. SF-103, Liege, Feb. 1971

MA- 48 Delcourt-Bon C. R. and Stenne L.,
Material Characterization by the Finite Element Method, *J. Reinf. Plast. Comp.*, Vol. *1*, 1982, 82-91

MA- 49 Delneste L. and Perez B.,
An Inelastic Finite Element Model of 3D Carbon-Carbon Composites, *AIAA J.*, Vol. *21*, No. 8, 1983, 1143-49

MA- 50 Ditcher A. K. and Webber J. P. H.,
Nonlinear Mechanical and Thermal Responses of a Unidirectional Carbon-Fiber-Reinforced Plastic, *J. Strain Anal. Engng. Design*, Vol. *14*, No. 4, 1979, 149-156

MA- 51 Dvorak G. J. and Bahei-El-Din Y. A.
Plasticity Analysis of Fibrous Composites, *J. Appl. Mechanics*, ASME, Vol. *49*, 1982, 327-335

MA- 52 Dvorak G. J. and Bahei-el-Din Y. A.,
Plasticity of Composite Lamintes, Proc. Res. Workshop Mech. Comp. Mat., Duke Univ., 1978, 32-54

MA- 53 Dvorak G. J. and Bahei-el-Din Y. A.,
Elastic-Plastic Behaviour of Fibrous Composites, *J. Mech. Phys. Solids*, Vol. *27*, No. 1, 1979, 51-72

MA- 54 Dvorak G. J. and Rao M. S.,
Plasticity Theory of Fibrous Composites - Part I, Duke Univ., Dept. of Civil Engng., Durham, 1973
MA- 55 Dvorak, G. J. and Rao, M. S.,
Yielding in Unidirectional Composites Under External Loads and Temperature Changes, J. Comp. Mat., Vol. 7, 1973, 194-216
MA- 56 Ebert L. J. and et al.,
Finite Element Analysis System for the Mechanical Behaviour of Oriented Fiber Composite Materials Under Combined Stresses, Case Western Reserve Univ., TR-75-42, Cleveland, 1974
MA- 57 Epstein M. and Glockner P. G.,
Nonlinear Analysis of Multilayered Shells, Solids Struct., Vol. 13, 1978, 1081-
MA- 58 Epstein M. and Huttelmaier H. P.,
A Finite Element Formulation for Multilayered and Thick Plates, Computers & Structures, Vol. 16, No. 5, 1983, 645-650
MA- 59 Ericson W. A.,
An Investigation of Initial Yield Surfaces for Unidirectional Reinforced Composites, Naval Postgr. School, Monterey, 1972
MA- 60 Figueiras J. A. and Owen D. R. J.,
Elasto-Plastic Analysis of Anisotropic Shell Structures, in Finite Elem. Softw. Plates Shells, Pineridge Press
MA- 61 Foye, R. L. and Baker, D. J.,
Design/Analysis Methods for Advanced Composite Structures, Rep. AFML-TR-70-299, 1971
MA- 62 Futakawa, A. and et al.,
Deformation and Stress of Antenna Made of Glass Fiber Reinforced Plastics Under Wind Force, 4th Int. Conf. Comp. Mat., To,yo, 1985, 1767-74
MA- 63 Garcia, R.,
An Experimental and Analytical Investigation of the Rail Shear-Test Method as Applied to Composite Materials, Exp. Mech., Vol. 20, No. 8, 1980, 272-279
MA- 64 Giavotto, V. and et al.,
Anisotropic Beam Theory and Applications, Computers & Structures, Vol. 16, 1-4, 1983, 403-413
MA- 65 Giavotto, V. and et al.,
Evaluation of Section Properties for Hollow Composite Beams, V. Europ. Rotorcraft Power Lift Aircr., Paper 35, 1979
MA- 66 Giavotto, V. and et al.,
R & D on Composite Rotor Blades at Agusta, VIII Eur Rotorcraft Power Lift Aircr., Paper 64, 1981
MA- 67 Griffin, O. H.,
Three-Dimensional Curing Stresses in Symmetric Cross-Ply Laminates with Temperature-Depenedent Properties, J. Comp. Mat., Vol. 17, 1983, 449-463
MA- 68 Griffin, O. H. and Roberts, J. C.,
Numerical/Experimental Correlation of Three-Dimensional Thermal Stress Distributions in Graphite/Epoxy Laminates, J. Comp. Mat., Vol. 17, 1983, 539-548
MA- 69 Griffin, O. H. and et al.,
Three Dimensional Inelastic Finite Element Analysis of Laminated Composites, J. Comp. Mat., Vol. 15, 1981, 543-560
MA- 70 Guess, T. R. and Haizlip, C. B.,
End-Grip Configurations for Axial Loading of Composite Tubes, Exp. Mech., Vol. 20, 1980, 31-36
MA- 71 Gupta, B. P. and Finney, R. H.,
Application of Finite-Element Method to the Analysis of High-Capacity Laminated Elastomeric Parts, Experimental Mech., March 1980, 103-108
MA- 72 Hale, J. M. and Ashton, J. N.,
A Novel Tensile Specimen for Sheet Material, Composite, Vol. 15, No. 1, 1984, 67
MA- 73 Hamada, S. and Arizumi, Y.,
Finite Element Analysis of Continuous Composite Beams with Incomplete Interaction, Proc. JSCE, No. 265, 1977, 1-9
MA- 74 Herakovich, C. T. and Itani, R. Y.,

Elastic-Plastic Torsion of Nonhomogeneous Bars, J. Engng. Mech. Div., ASCE, EM5, Oct. 1976, 757-769
MA- 75 Herakovich, C. T. and et al.,
Finite Element Analysis of Mechanical and Thermal Edge Effects in Composite Laminates, in Comp. Mat., Army Symp. Solid Mech., Cape Code, MA, 1976
MA- 76 Herakovich, C. T. and et al.,
A Comparative Study of Composite Shear Specimens Using the Finite Element Method, in Test Meth. Design Allow Fibr. Comp., ASTM STP734, 1981, 129-151
MA- 77 Herrmann, K. and Mihovsky, I.,
Plastic Behaviour of Fibre-Reinforced Composites and Fracture Effects, 4th Nat. Cong. Theor. Appl. Mech., Varna, Bulgaria, Sept. 1981, 431-436
MA- 78 Herrmann, I. R.,
Finite Elemenet Modelling of Composite Edge Effects, 7th Conf. Electr. Comp., asce, 1979, 593-607
MA- 79 Hirst, M. J. S. and Yeo, M. F.,
The Analysis of Composite Beams Using Standard Finite Element Programs, *Computers & Structures*, Vol. *11*, 1980, 233-237
MA- 80 Hong, C. S. and Crews, J. H.,
Stress Concentration Factors for Finite Orthotropic Laminates With a Circular Hole and Uniaxial Loading, NASA, Paper 1469, 1979
MA- 81 Hudspith, I. and Duthie, T.,
Modelling Composite Honeycomb Structures, ASAS Shell Anal. Sem., London, 1980
MA- 82 Hyman, B. I. and Nilforoush, J. M.,
Effect of Fibre Distribution on Stress and Strain Concentration at Holes in Composite Plates, 6th Symp. Comp. Mat. Engng. Design, Washington Un., 1972, 621-35
MA- 83 Idelsohn S. and et al.,
Pre- and Post-Degradation Anaysis of Composite Materials with Different Moduli in Tension and Compression, *Comp. Meth. Appl. Mech. Engng.*, Vol. *30*, No. 2, 1982, 133-149
MA- 84 Immen, F. H. and Foye, R. L.,
New Insights in Structural Design of Composite Rotor Blades for Helicopters, 4th Int. Conf. Comp. Mat., Tokyo, Oct. 1982, 1673-84
MA- 85 Ishakian, V. G. and Hollaway, L.,
Application of the Finite Element Method to the Analysis of a Skeletal/Continuum GRP Space Structure, *Composites*, Vol. *10*, 1979, 81-88
MA- 86 Jones, R. and Callinon, R. J.,
Analysis of Composite Laminates and Fibre Composite Repair Schemes, *Fibre Sci. Tech.*, Vol. *15*, No. 3, 1981, 199-207
MA- 87 Jones, R. and et al.,
Analysis of Multi-Layer Laminates Using Three-Dimensional Super-Elements, *Int. J. Num. Meth. Engng.*, Vol. *20*, 1984, 583-587
MA- 88 Jones, R. M. and Nelson, A. R. N.,
A New Material Model for the Nonlinear Biaxial Behavour of ATJ-S Graphite, *J. Comp. Mat.*, Vol. *9*, 1975, 10-27
MA- 89 Kang, S. I.,
On the Determination of Effective Moduli of Composite Materials by a Three-Dimensional Finite Element Method, Ph D Thesis, Georgia Inst. of Tech., Atlanta, 1974
MA- 90 Khot, N. S. and Venkayya, V. B.,
Optimum Design of Advanced Composite Structures for Static Loads, 6th Symp. Comp. Mat. Eng. Design, St. Louis, May 1972, 417-428
MA- 91 Khot, N. S. and Venkayya, V. B.,
Application of Optimally Criterion to Fiber-Reinforced Composites, Rep. AFFDL-TR-73-6, May 1973
MA- 92 Kikukawa, H. and Takaki, J.,
Experimental Investigation of Anisotropic Laminate Structural Behaviour, 4th Int. Conf. Comp. Mat., Tokyo, Oct. 1982, 405-412
MA- 93 Kondo, K. and Aoki, T.,
Longitudinal Shear Modulus of Unidirectional Composites, 4th Int. Conf. Comp. Mat., Tokyo, Octo. 1982, 357-364
MA- 94 Kowalski, M. F. and et al.,

Design Analysis of Automotive Composite Structures, 3rd Int. Conf. Vehicle Str. Mech., Troy, Oct. 1979, 83-90
MA- 95 Kulkarni, H. and Beardmore, P.,
Design Methodology for Automotive Components Using Continuous Fibre-Reinforced Composite Materials, *Composites*, Vol. *11*, No. 4, 1980, 225-235
MA- 96 Kuppusamy, T. and Reddy, J. N.,
A Three-Dimensional Nonlinear Analysis of Cross-Ply Rectangular Composite Plates, *Computers & Structures*, Vol. *18*, No. 2, 1984, 263-272
MA- 97 Kwok, W. L. and Cheung, Y. K.,
Analysis of Circular and Annular Laminated Thick Plates, in Finite Elem. Meth. Engng., Univ. NSW, Australia, 1974, 177-194
MA- 98 Lakshmikantham, C. and Tong, P.,
Stresses Around Holes in Stiffened Composite Panels Using Laurent-Series and Finite Element Methods, 5th Int. Conf. Exp. Stress An., Udine, 1974
MA- 99 Lakshminarayana, H.,
Finite Element Analysis of Laminated Composite Shell Structures - Part 1: Static Analysis, NAL-TM-ST-403/170-77, Bangalore, India, Sept. 1977
MA- 100 Lakshminarayana, H. and Sridharamurthy, S.,
Vibration of Laminated Composite Shells of Revolution - A Finite Element Analysis, NAL-TM-ST-405/198-78, Bangalore, India, Nov. 1978
MA- 101 Lakshminarayana, H. and Sundareshan, M. J.,
Stress Concentration Around Circular Cutouts in Laminated Composite Cylindrical Shells: Experimental Investigations and Finite Element Analysis, *J. Aeron. Soc. India*, Vol. *35*, 1983, 13-22
MA- 102 Lakshminarayana, H. and Viswanalt, S.,
Application of the Finite Element Method to the Analysis of Laminated Composite Shell Structures, Symp. Appl. Comp. Meth- Engng., Un. South CA, L. Angeles, Aug. 1977
MA- 103 Lakshminarayana, H. and Viswanath, S.,
Influence of Material Orthotropy on the State of Stress Around a Circular Hole in a Cylindrical Shell, J. Press Vess. Tech., ASME, Vol. *99*, 1977
MA- 104 Lakshminarayana, H. V. and Viswanath, S.,
A Correlation Study of Finite-Element Modelling for the Stress Analysis of Composite-Material Laminates, *J. Strain Anal. Engng. Design*, Vol. *13*, No. 4, 1978, 205-212
MA- 105 Larder, R. A.,
The Stochastic Finite Element Simulation of the Nonlinear Structural Response of Fibrous Composite Materials, Dr. Eng. Thesis, U. C. Davis, 1975
MA- 106 Larder, R. A. and Beadle, C. W.,
The Stochastic Finite Element Simulation of Parallel Fibre Composites, *J. Comp. Mat.*, Vol. *10*, 1976, 21-31
MA- 107 Lee, J. D.,
Three Dimensional Finite Element Analysis of Layered Fiber-Reinforced Composite Materials, *Computers & Structures*, Vol. *12*, 1980, 319-339
MA- 108 Lee, T. Y.,
Structural Analysis of Filament-Winding Pressure Vessels, 4th Int. Conf. Comp. Mat., Tokyo, Oct. 1982, 421
MA- 109 Lin, F. T.,
The Finite Element Analysis of Laminated Composites, Ph D Diss, Virginia Polyt. Inst. State Un., Blacksburg, 1972
MA- 110 Lin, T. H. and Salinas, D.,
Initial Yield Surface of a Unidirectionally Reinforced Composite, ASME Paper, 71-APMW-19, 1971
MA- 111 Lin, T. H. and *et al.*,
Elastic-Plastic Analysis of Unidirectional Composites, *J. Comp. Mat.*, Vol. *6*, 1972
MA- 112 Marloff, R. H. and Gabrielse, S. E.,
Further Evaluation of Proposed Biaxial Stress Test Specimen for Composite Materials, 3rd Int. Conf. Comp. Mater., Paris, 1980
MA- 113 Martin, C. W. and Zitek, S. J.,
Geometric Nonlinearity in Composite Pressure Vessels, ADINA Conf., MIT, Cambridge, Aug., 1977, 189-204

MA- 114 Mau, S. T. and et al.,
Finite Element Solutions for Laminated Thick Plates, *J. Comp. Mat.*, Vol. *6*, 1972, 304-311
MA- 115 Mawenya, A. S. and Davies, J. D.,
Finite Element Bending Analysis of Multilayer Plates, *Int. J. Num. Meth. Engng.*, Vol. *8*, 1974, 215-225
MA- 116 McCullers, L. A. and Naberhaus, J. D.,
Automated Structural Design and Analysis of Advanced Composite Wing Models, *Computers & Structures*, Vol. *3*, No. 4, 1973, 925-935
MA- 117 McLay, R. W. and Murphy, M. C.,
An Energy Search Technique for Composite Fracture, in Interdiscip. Finite Elem. Anal., Cornell Univ., 1981, 119-135
MA- 118 Monforton, G. R.,
Stiffness Matrix for Sandwich Beams With Thick Anisotropic Laminated Faces, *Computers & Structures*, Vol. *10*, 1979, 547-551
MA- 119 Monforton, G. R. and Schmit, L. A.,
Finite Element Analysis of Sandwich Plates and Cylindrical Shells With Laminated Faces, Proc. 2nd Conf. Matrix Meth. Str. Mech., WPAFB, Ohio, 1969
MA- 120 Monib, M. M. and Adams, D. F.,
Nonlinear Three-Dimensional Finite Element Analysis of Composite Laminates, Univ. of Wyoming, UWME-DR-OO1-102-1, Nov. 1970
MA- 121 Murthy, P. V. R. and Rao, K. P.,
Analysis of Curved Laminated Beams of Bimodulus Composite Materials, *J. Comp. Mat.*, Vol. *17*, 1973, 435-448
MA- 122 Murthy, P. V. R. and Rao, K. P.,
Finite Element Analysis of Laminated Anisotropic Beams of Bimodulus Materials, *Computers & Structures*, Vol. *18*, No. 5, 1974, 779-787
MA- 123 Murthy, P. V. R. and et al.,
Finite Element Analysis of Laminated Anisotropic Curved Hollow Bars, Indian Inst. of Sci., Rep. AE 370S, Bangalore, Dec. 1982
MA- 124 Nameth, M. P. and et al.,
On the Off-Axis Tension Test for Unidirectional Composites, Cement, Concr. & Aggreg., ASTM, 1983, 61-68
MA- 125 Neu, T. F.,
Finite Element Analysis of Edge Effects in Angle-Ply Composite Laminates, Inster., PA, NADC-74051-30, 1974
MA- 126 Nickell, R. E. and Sato, T.,
Finite Element Stress Analysis of Orthotropic Layered Shells of Revolution Using a Curved Shell Element, Redstone Res. Lab., Rep. S-264, Huntsville, Oct. 1970
MA- 127 Nishioka, T. and Atluri, S. N.,
Stress Analysis of Holes in Angle-Ply Laminates: An Efficient Assumed Stress 'Special-Hole-Element' Approach and a Simple Estimation Method, *Computers & Structures*, Vol. *15*, No. 2, 1982, 135-147
MA- 128 Noor, A. K. and Mathers, M. D.,
Nonlinear Finite Element Analysis of Laminated Composite Shells, Proc. Conf. Comp. Meth. Nonlin. Mech., Austin, 1974
MA- 129 Noor, A. K. and Mathers, M. D.,
Shear-Flexible Finite-Element Models of Laminated Composite Plates and Shell, NASA TN D-8044, Dec. 1975
MA- 130 Noor, A. K. and Mathers, M. D.,
Finite Element Analysis of Anisotropic Plates, *Int. J. Num. Meth. Engng.*, Vol. *11*, 1977, 289-307
MA- 131 Noor, A. K. and Mathers, M. D.,
Anisotropy and Shear Deformation in Laminated Composite Plates, *AIAA J.*, Vol. *14*, 1976, 282-285
MA- 132 Nyssen, C.,
Modelisation due comportement mecanique des resines armees, LTAS Rep. SF-82, Liege, Aug. 1978
MA- 133 Nyssen, C.,
Modelisation par elements finis du comportement non lineaire de structures aerospatiales, Doctoral Thesis, LTAS Rep. SF-86, Liege, Dec. 1978

MA- 134 Nyssen, C. and Beckers, P.,
Finite Element Linear and Nonlinear Analysis of Composite Materials, Proc.
ADINA Meet., MIT, Cambridge, Aug. 1979
MA- 135 O'Hara, G. P.,
Finite Element Analysis of Fiber Wrapped Shells with Non-Symmetric Loads,
6th NASTRAN Coll., NASA Publ. 2018, Oct. 1977, 369-381
MA- 136 Ostrowski, J. and et al.,
Poisson's Stresses in Fibre Composites. I: Analysis, *J. Strain Anal.*, Vol. *19*,
No. 1, 1984, 43-49
MA- 137 Owen, D. R. J. and Figueiras, J. A.,
Elasto-Plastic Analysis of Anisotropic Plates and Shells by the Semiloof
Element, *Int. J. Num. Meth. Engng.*, Vol. *19*, 1983, 521-539
MA- 138 Owen, D. R. J. and Figueiras, J. A.,
Anisotropic Elasto-Plastic Finite Element Analysis of Thick and Thin Plates
and Shells, *Int. J. Num. Meth. Engng.*, Vol. *19*, 1983, 541-566
MA- 139 Owen, D. R. J. and Prakash, A.,
Finite Element Analysis of Non-Linear Composite Materials by Use of Overlay
Systems, *Computers & Structures*, Vol. *4*, No. 6, 1974, 1251-67
MA- 140 Pack, S. C.,
Finite Element Analysis of Fibrous Composite Materials, Master Thesis, Syracuse
Univ., 1979
MA- 141 Panda, S. and Natarajan, R.,
Analysis of Laminated Composite Shell Structures by Finite Element Method,
Computers & Structures, Vol. *14*, 3-4, 1981, 225-230
MA- 142 Panda, S. and Natarajan, R.,
Finite Element Analysis of Laminated Shells of Revolution, *Computers & Structures*, Vol. *6*, 1976, 61-64
MA- 143 Panda, S. C. and Natarajan, R.,
Finite Element Analysis of Laminated Composite Plates, Int. J. Num. Meth.
Engng., Vol. *14*, 1979, 69-79
MA- 144 Parhizgar, S. and et al.,
Application of the Concept of Strain Energy Density to Unidirectional Fibre
Reinforced Composites, Int. Symp. Mech. Behav. Struct. Media, Ottawa, May 1981
MA- 145 Parton, G. M. and Shendy-El-Barbary,
A Finite Element Analysis for Cement Composite Sandwich Plates, *Int. J. Cement Comp. Lightweight Concr.*, Vol. *5*, No. 3, 1983, 181-191
MA- 146 Patel, H. P. and Kennedy, R. H.
Nonlinear Finite Element Analysis for Composite Structures of Axisymmetric
Geometry and Loading, *Computers & Structures*, Vol. *15*, No. 1, 1982, 79-84
MA- 147 Pian, T. H. H.,
Interactive Program in Design and Analysis of Composite Materials, MIT, Rep.
asrl-TR-163-1, 1974
MA- 148 Porowski, J. S. and O'Donnell, W. J.,
Plastic Strain Concentrations in Ligaments, J. Press Vess. Tech., ASME, May
1977
MA- 149 Pryor, C. W.,
Finite Element Analysis of Laminated Anisotropic Plates Including Transverse
Shear Deformation, Ph D Thesis, Virginia Polytech. Inst., 1970
MA- 150 Pryor, C. W. and Barker, R. M.,
A Finite Element Analysis Including Transverse Shear Effects for Application
to Laminated Plates, *AIAA J.*, Vol. *9*, 1971, 912-917
MA- 151 Pryor, C. W. and Barker, R. M.,
Finite Element Analysis of Bending-Extensional Coupling in Laminated Composites,
J. Com. Mat., Vol. *4*, 1970, 549-552
MA- 152 Purdy, D. M.,
Discrete Element Analysis of Composite Structures, *Comp. Mat.*, Vol. *8*, 1975,
1-31
MA- 153 Purushothaman, P. and et al.,
Studies on Two Phase Composite Materials Through Computer Simulation, in
Theory Pract. FE Str. Anal., Tokyo Pr., Tokyo, 1973, 233-245
MA- 154 Putcha, N. S. and Reddy, J. N.,
Three Dimensional Finite-Element Analysis of Layered Composite Plates, Adv.

Aerosp. Str., Winter Ann. Meet., ASME, Phoenix, 1982
MA- 155 Raju, I. S. and Crews, J. H.
Polar Symmetry in Three-Dimensional Analysis of Laminates With Angle-Plies, NASA Tech. Mem., TM-81852, July 1980
MA- 156 Raju, I. S. and Crews, J. H.,
Three-Dimensional Analysis of /0/90/s and /90/0/s Laminates with a Central Circular Hole, *Comp. Technol. Review*, Vol. 4, No. 4, 1982, 116-124
MA- 157 Raju, I. S. and et al.,
A New Look at Numerical Analyses of Free-Edge Stresses in Composite Laminate, NASA TP- 1751, Dec. 1980
MA- 158 Ramanamurthy, P. V. and et al.,
Finite Element Analysis of Laminated Anisotropic Curved Hollow Bars, Dept. of Aeronaut. Rep. AE 370S, Bangalore, 1982
MA- 159 Rao, M. S. and Dvorak, G. J.,
Plasticity Theory of Fibrous Composites - Part II, Duke Univ., Dept. of Civil Engng., Durham, 1973
MA- 160 Rashid, Y. R.,
Finite Element Analysis of Axisymmetric Composite Structures, AEC R&D, Rep. ATO4-3-167, 1965
MA- 161 Rashid, Y. R.,
Analysis of Axisymmetric Composite Structures by the Finite Element Method, Nuclear Engng. Design, Vol. 3, 1966, 163-182
MA- 162 Reddy, J. N.,
Survey of Recent Research in the Analysis of Composite Plates, *Comp. Tech. Review*, Vol. 4, No. 3, 1982, 101-104
MA- 163 Redy, J. N.,
Finite Element Modelling of Layered Anisotropic Composite Plates and Shells: A Review of Recent Research, *Shock and Vib. Digest*, Vol. 13, No. 12, 1981, 3-12
MA- 164 Reddy, J. N.,
Analysis of Layered Composite Plates Accounting for Large Deflections and Transverse Shear Strains, in Rec. Adv. Non-Lin. Comp. Mech., Pineridge, Swansea, 1982, 155-202
MA- 165 Reddy, J. N. and Bert, C. W.,
Analysis of Plates Constructed of Fiber-Reinforced Bimodulus Material, Mech. Bimodulus Mat., ASME Winter Ann. M., New York, 1979
MA- 166 Reddy, J. N. and Chao, W. C.,
Nonlinear Oscillations of Laminated, Anisotropic, Rectangular Plates, J. Appl. Mech., ASME, Vol. 49, No. 2, 1982, 396-402
MA- 167 Reddy, N. J. and Chao, W. C.,
Nonlinear Bending of Bimodular-Material Plates, *Int. J. Solids Str.*, Vol. 19, No. 3, 1983, 229-237
MA- 168 Reddy, J. N. and Chao, W. C.,
A Comparison of Closed-Form and Finite Element Solutions of Thick, Laminated Anisotropic Rectangular Plates, Nuclear Engng. Design, Vol. 61, 1981, 153-167
MA- 169 Reddy, J. N. and Chao, W. C.,
Finite Element Analysis of Laminated Bimodulus Composite-Material Plates, *Computers & Structures*, Vol. 12, 1980, 245-251
MA- 170 Reddy, J. N. and Chao, W. C.,
Non-Linear Bending of Thick Rectangular, Laminated Composite Plates, *Int. J. Non-Linear Mech.*, Vol. 16, 1982, 291-301
MA- 171 Renieri, G. D. and Herakovich, C. T.,
Nonlinear Analysis of Laminated Fibrous Composites, NASA CR-148317, June 1976
MA- 172 Rich, M. J. and Lowry, D. W.,
Design Analysis and Test of Composite Curved Frames for Helicopter Fuselage Structure, 24th Str., Str. Dyn. Mater. Conf., Lake Tahoe, May 1983, 730-737
MA- 173 Rizzo, R. R. and Vicario, A. A.,
A Finite Element Analysis of Laminated Anisotropic Tubes, *J. Comp. Mat.*, Vol. 4, 1970, 344-359
MA- 174 Rohwer, K.,
On the Determination of Edge Stresses in Layered Composites, *Nuclear Engng. Design*, Vol. 70, 1982, 57-65

MA- 175 Rohwer, K.,
Stresses and Deformations in Laminated Test Specimens of Carbon Fiber
Reinforced Composites, DFVLR-Forschungsber 82-15, Braunschweig, Germany, 1982
MA- 176 Ross, A. L.,
Designing with Three Directional Composites, ASME Meet., 74-DE-25, 1974
MA- 177 Rybicki, E. E. and Schmueser, D. W.,
Effect of Stacking Sequence and Lay-Up Angle on Free Edge Stresses Around a
Hole in a Laminated Plate Under Tension, *J. Comp. Mat.*, Vol. *12*, 1978, 300-313
MA- 178 Rybicki, E. F.,
Approximate Three-Dimensional Solutions for Symmetric Laminates Under In-Plane
Loading, *J. Comp. Mat.*, Vol. *5*, 1971, 354-360
MA- 179 Rybicki, E. F. and Schmueser, D. W.,
Three Dimensional Finite Element Stress Analysis of Laminated Plates Containing
a Circular Hole, Battelle Columbus Lab., AFML-TR-76-92, Aug. 1976
MA- 180 Sander, G. and Nyssen, C.,
Modelisation des materiaux composites dans les methods d'elements finis, XIV
Int. Cong. int. Aeronaut, Paris, Juin 1979
MA- 181 Sander, G. and et al.,
Modelisation par elements finis de certains comportements non lineaire des
materiaux composites, Meth. Num. dans les Scien., GAMNI, Paris, 1980, 607-622
MA- 182 Schaffer, B. G. and Adams, D. F.,
Nonlinear Viscoelastic Analysis of a Unidrectional Composite Material, J.
Appl. Mech., Vol. *48*, No. 4, 1981, 859-865
MA- 183 Schmit, L. A. and Monforton, G. R.,
Finite Deflection Discrete Element Analysis of Sandwich Plates and Cylindrical
Shells with Laminated Faces, AIAA/ASME 10th Str., Str. Dyn. Mater. Conf., New
Orleans, 1969
MA- 184 Schneermann, M. W.,
Design, Analysis and Testing of Two Concepts for a Dimensional Stable Structure,
4th Int. Conf. Comp. Mat., Tokyo, Octo. 1982, 1693-99
MA- 185 Seide, P. and Chang, P. N.,
Finite Element Analysis of Laminated Plates and Shells, NASA CR-157107, 1978
MA- 186 Snyder, M. D. and Cruse, T. A.,
Crack Tip Stress Intensity Factors in Finite Anisotropic Plates, AFML-TR-73-209,
WPAFB, Ohio, 1973
MA- 187 Spilker, R. L.,
A Finite Element Model for Laminated Plates Including Transverse Shear
Deformation, ASRL TR 169-1, MIT, 1972
MA- 188 Spilker, R. L.,
A Hybrid-Stress Finite-Element Formulation for Thick Multilayer Laminates,
Computers & Structures, Vol. *11*, 1980, 507-514
MA- 189 Spilker, R. L. and Chou, S. C.,
Edge Effects in Symmetric Composite Lamintes: Importance of Satisfying the
Traction-Free-Edge Condition, *J. Comp. Mat.*, Vol. *14*, 1980, 2-20
MA- 190 Spilker, R. L. and Chou, S. C.,
Evaluation of a Hybrid-Stress Formulation for Thick Multilayer Laminates, 4th
Conf. Fibr. Comp. Str. Design, San Diego, Nov. 1978
MA- 191 Spilker, R. L. and Munir, N. I.,
Comparison of Hybrid-Stress Element Through-Thickness Distributions
Corresponding to a High-Order Plate Theory, *Computers & Structures*, Vol. *11*,
1980
MA- 192 Spilker, R. L. and et al.,
Use of the Hybrid Stress Finite Element Model for the Static and Dynamic
Analysis of Multilayer Composite Plates and Shells, Army Mat. Mech. Res. Center,
asrl-TR-181-2, Watertown, Sept. 1976
MA- 193 Starnes, J. H. and Haftka, R. T.,
Preliminary Design of Composite Wings for Buckling - Strength and Dispalcement
Constraints, *J. Aircraft*, 1979
MA- 194 Sunder, P. J. and et al.,
Finite Element Analysis of 3-Ply Laminated Conical Shell for Flutter, Int. J.
Num. Meth. Engng., Vol. *19*, 1983, 1183-92
MA- 195 Tabaddor, F.,

Two-Dimensional Finite Element Analysis of Bi-Modulus Material, *Fibre Sci. Technol.*, Vol. *14*, No. 3, 1981, 229-240
MA- 196 Tabaddor, F.,
The Finite Element Technique in the Study of Composite Materials, in Devel. Comp. Mat., Applied Sci. Publ., Essex, 1981
MA- 197 Tabaddor, F.,
Large Deformations of Cord-Reinforced Multilayered Shells, *J. Fibre Sci. Technol.*, Vol. *12*, 1979, 253-267
MA- 198 Teh, K. K. and et al.,
On the Three Dimensional Analysis of Thick Laminated Plates, 4th Int. Conf. Austral. FEM, Melbourne, Aug. 1982, 169-173
MA- 199 Tian, X. Y. and Wang, T. K.,
An Analysis of Nonlinear Instability of Reinforced Composite Plates, Acta Materiae Compositae Sinica, Vol. *1*, No. 1, 1984
MA- 200 Uitto, R. J.,
Finite Element Analysis in Composite Structures,
ANSYS Conf. Exhib., Pittsburgh, April 1983
MA- 201 Venkatesh, A. and Rao, K. P.,
Analysis of Laminated Shells with Laminated Stiffeners Using Rectangular Finite Elements, *Comp. Meth. Appl. Mech. Engng.*, Vol. *38*, 1983, 255-272
MA- 202 Vidouse, F.,
Determination of the Technical Constants of Laminates in Oblique Directions, NASA Tech. Mem., TM-75719, Oct. 1979
MA- 203 Waddoups, M. E. and et al.,
Composite Wing for Transonic Improvement - Advanced Analysis Evaluation, Convair Aerospace Div., Fort Worth, Nov. 1971
MA- 204 Wang, A. S. D. and Crossman, F. W.,
Some New Results on Edge Effects in Symmetric Composite Laminates, *J. Comp. Mat.*, Vol. *11*, No. 1, 1977, 92-106
MA- 205 Wang, A. S. D. and Crossman, F. W.,
Edge Effects on Thermally Induced Stresses in Composite Laminates, *J. Comp. Mat.*, Vol. *11*, 1977, 300-312
MA- 206 Wang, S. S. and Yuan, F. G.,
A Singular Hybrid Finite Element Analysis of Boundary-Layer Stresses in Composite Laminates, *Int. J. Solids Str.*, Vol. *19*, No. 9, 1983, 825-837
MA- 207 Wang, S. S. and et al.,
A Mixed-Mode Crack Analysis of Rectilinear Anisotropic Solids Using Conservation Laws of Elasticity, *Int. J. Fract.*, Vol. *16*, No. 3, 1980, 247-259
MA- 208 Wang, T. K. and Ling, D. F.,
Stability of Composite Laminated Plates by Finite Element Method, 2nd Nat. Conf. Comp. Mater., Beijing, China, 1981
MA- 209 Wang, T. K. and Luo, L. S.,
Finite Element Analysis of Composite Laminated Plates and Cylindrical Shells with Holes, 3rd Nat. Conf. Comp. Mater., Hangzhou, China, Nov. 1984
MA- 210 Weeks, G. E. and Cost, T. L.,
Complex Stress Response and Reliability Analysis of a Composite Elastici- Viscoelastic Missile Configuration Using Finite Elements, Mech. Res. Comm., Vol. *7*, No. 2, 1980, 59-63
MA- 211 Wegmuller, A. W. and Amer. H. N.,
Nonlinear Response of Composite Steel-Concrete Bridges, *Computers & Structures*, Vol. *7*, 1977, 161-169
MA- 212 Weinstein, F. and et al.,
Thermoelastic Stress Analysis of Anisotropic Composite Sandwich Plates by Finite Element Method, *Computers & Structures*, Vol. *17*, No. 1, 1983, 31-36
MA- 213 Weisshaar, T. A. and Garcia, R.,
Analysis of Graphite/Polyimide Rail Shear Specimens Subjected to Mechanical and Thermal Loading, NASA-CR-157769, Oct. 1978
MA- 214 Whang, B.,
A Finite Element Analysis of Laminated Orthotropic Plates and Shallow Shells, NSRDC, AD-704495, March 1970
MA- 215 Whang, B.,
Laminated Orthotropic Plates and Shallow Shells, *Computers & Structures*, Vol. *1*,

No. 3, 1971, 465-493
MA- 216 Whitcomb, J. D. and Raju, I. S.,
Superposition Method for Analysis of Free-Edge Stresses, J. Comp. Mat., Vol. 17, 1983, 492-507
MA- 217 Whitcomb, J. D. and et al.,
Reliability of the Finite Element Method for Calculating Free Edge Stresses in Composite Laminates, Computers & Structures, Vol. 15, No. 1, 1982, 23-37
MA- 218 Whiteside, J. B. D.,
The Behaviour of Advanced Filamentary Composite Plates With Cutouts, IIT Res. Inst., Chicago, June 1973
MA- 219 Wilt, J. C.,
Use of Composite Materials in NASTRAN, 24th Str., Str. Dyn. Mater. Conf., Lake Tahoe, May 1983, 612-622
MA- 220 Witt, W. and Palazotto, A. N.,
Nonlinear Analysis of Laminated Composite Plates, Int. Symp. Mech. Behav. struct. Media, Ottawa, May 1981, 401-407
MA- 221 Woerndle, R.,
Calculation of the Cross Section Properties and the Shear Stresses and Composite Rotor Blade, VII Europ. Rotorcraft Power Lift Aircr., Paper 65, 1981
MA- 222 Woolstencroft, D. H. and et al.,
A Comparison of Test Techniques Used for the Evaluation of the Unidirectional Compressive Strength of Carbon Fibre-Reinforced Plastic, Composites, Vol. 12, No. 4, 1981, 275-280
MA- 223 Woolstencroft, D. H. and et al.,
The Compressive Behaviour of Carbon Fibre Reinforced Plastic, 4th Int. Conf. Comp. Mat., Tokyo, Oct. 1982, 439-446
MA- 224 Yamada, Y. and et al.,
A Finite Element Simulation of Mechanical Properties of Composite Materials, Comp. Mat. Str., Fukugo Zairyo Kenkyu, Vol. 3, No. 4, 1974, 7-11
MA- 225 Yamada, Y. and et al.,
Finite Element Analysis of Composite Material Behaviours From Mechanical Properties of Constituent Materials, 4th Nat. Symp. Matrix Meth. Str. Design, Japan, 1975, 471-477
MA- 226 Yasuda, M. and et al.,
The Compressive Behaviour of a Carbon Fibre-Reinforced Plastic Ring with a Non-Linear Section, Composites, Vol. 15, No. 2, 1984, 134-138
MA- 227 Yettram, A. L.,
Use of Finite Element Method for Investigating the Thermal Expansion of Composites, Fibre Sci. Tech., Vol. 10, No. 3, 1977, 233-237
MA- 228 Zadorozny, E. A.,
Space Shuttle Carbon-Carbon Composite Hot Structure, 24th Str., Str. Dyn. Mater. Conf., Lake Tahoe, May 1983, 359-368
MA- 229 Zako, M. and Miyoshi, T.,
Study on Dispersion of Strength for Composite Materials, J. Soc. Mat. Sci., Japan, Vol. 28, 1979, 33-39
MA- 230 Zybert, J. J. and et al.,
Theoretical Determination of Composite Material Elastic Compliances, Composites, Vol. 11, No. 3, 1980, 175-180

Micromechanical Analysis (MI)

MI- 1 Adams, D. F.,
Practical Problems Associated with the Application of the Finite Element Method to Composite Material Micromecha-ical Analyses, Fibre Sci. Tech., Vol. 7, No. 2, 1974, 111
MI- 2 Adams, D. F.,
A Micromechanical Analysis of Crack Propagation in an Elastoplastic Composite Material, Fibre Sci. Tech., Vol. 7, No. 4, 1974, 237
MI- 3 Adams, D. F. and Crane, D. A.,
Combined Loading Micromechanical Analysis of a Unidirectional Composite, Composites, Vol. 15, No. 3, 1984, 181-192

MI- 4 Adams, D. F. and Crane, D. A.,
Finite Element Micromechanical Analysis of a Unidirectional Composite
Including Longitudinal Shear Loading, *Computers & Structures*, Vol. *18*, No. 6,
1984, 1153-65

MI- 5 Adams, D. F. and Miller, A. K.,
Hygrothermal Microstresses in a Unidirectional Composite Exhibiting Inelastic
Material Behaviour, *J. Composite Mat.*, Vol. *11*, 1977, 285-299

MI- 6 Agarwal, B. D.,
Micromechanics Analysis of Composite Materials Using Finite Element Methods,
Illinois Inst. of Tech., Chicago, 1972

MI- 7 Agarwal, B. D. and Bansal, R. K.,
Plastic Analysis of Fibre Interactions in Discontinuous Fibre Composites,
Fibre. Sci. Tech., Vol. *10*, No. 4, 1977, 281-297

MI- 8 Agarwal, B. D. and Broutman, L. J.,
Three Dimensional Finite Element Analysis of Spherical Particle Composites,
Fibre Sci. Tech., Vol. *7*, No. 1, 1974, 63

MI- 9 Akbarzadeh, A. and Adams, D. F.,
A Hybrid Finite Element Micromechanical Analysis of Composite Materials,
Fibre Sci. Tech., Vol. *9*, No. 4, 1976, 277-295

MI- 10 Braun, H. and Herrmann, K.,
Analysis of Thermal Cracking of Unidirectionally Reinforced Composite
Structures in the Micromechanical Range, Proc. 5th Int. Conf. Fract., ICF 5,
Pergamon Press, 1981, 485-493

MI- 11 Broutman, L. J. and Agarwal, B. D.,
Theoretical Study of the Effect of the Interface on Composite Toughness, 28th
Annual Conf. Spi. Reinf. Plast. Div., Washington, D.C. 1973

MI- 12 Broutman, L. J. and Panizza, G.,
Micro-Mechanics Studies of Rubber-Reinforced Glassy Polymers, *J. Polymer Mat.*,
Vol. *1*, 1971, 95-109

MI- 13 Cairns, D. S. and Adams, D. F.,
Moisture and Thermal Expansion Properties of Unidirectional Composite Materials
and the Epoxy Matrix, *J. Reinf. Plast. Comp.*, Vol. *2*, 1983, 239-255

MI- 14 Carlsson, L.,
Interlaminar Stresses at a Hole in a Composite Member Subjected to In-Plane
Loading, *J. Comp. Mat.*, Vol. *17*, No. 3, 1983, 238-249

MI- 15 Chamis C. C. and Williams, G. C.,
Interply Layer Degradation Effects on Composite Structural Response, NASA Tech.
Mem. 83702, 1984

MI- 16 Crane, D. A. and Adams, D. F.,
Finite Element Micromechanical Analysis of a Unidirectional Composite Including
Longitudinal Shear Loading, Univ. of Wyoming, Rep. DR-101-101-1, Laramie, 1981

MI- 17 Crossman, F. W. and Wang, A. S. D.,
Stress Field Induced by Transient Moisture Sorption in Finite-Width Composite
Laminates, *J. Comp. Mat.*, Vol. *12*, 1978, 2-18

MI- 18 Farley, G. L. and Herakovich, C. T.,
Influence of Two-Dimensional Hygrothermal Gradients on Interlaminar Stresses
Near Free Edges, in Adv. Comp. Mater-Environ. Effects, ASTM, STP 658, 1978

MI- 19 Flynn, P. L. and Ebert, L. J.,
An Analytical Method for Predicting Plastic Flow in Notched Fiber Composite
Materials, *Metallurg. Trans.*, Vol. *8A*, May 1977, 723-728

MI- 20 Foye, R. L.,
Inelastic Micromechanics of Curing Stresses in Composites, *Inelastic Behav.
Comp. Mat.*, ASME, AMD, Vol. *13*, 1975

MI- 21 Foye, R. L.,
Theoretical Post-Yielding Behaviour of Composite Laminates, Part I - Inelastic
Micromechanics, *J. Comp. Mat.*, Vol. *7*, 1973, 178-193

MI- 22 Givler, R. C. and *et al.*,
Numerical Prediction of Fibre Orientation in Dilute Suspensions, *J. Comp. Mat.*,
Vol. *17*, No. 4, 1983, 330-343

MI- 23 Hayashi, I. and Iwamoto, J.,
Stress Distribution Near a Broken Point of Filament in Unidirectionally Fiber
Reinforced Composites, 17th Japan Cong. Mat. Res., Tokyo, 1973, 162-166

MI- 24 Herakovich, C. T.,
On the Relationship Between Engineering Properties and Delamination of Composite Materials, J. Comp. Mat., Vol. 15, 1981, 336-348

MI- 25 Herrmann, K. and Mattheck, C.,
Thermal Stresses in the Unit Cell of a Fiber-Reinforced Material, J. Thermal Stress, Vol. 2, 1979, 15-24

MI- 26 Hirai, T. and et al.,
Three-Dimensional Interlaminar Stress Distribution in Symmetric Composite Laminate, 4th Int. Conf. Comp. Mat., Tokyo, Oct. 1982, 243-250

MI- 27 Isakson, G. and Levy, A.,
Finite Element Analysis of Interlaminar Shear in Fibrous Composites, J. Comp. Mat., Vol. 5, No. 2, 1971, 273-276

MI- 28 Ko, W. L.,
Finite Element Microscopic Stress Analysis of Cracked Composite Systems, J. Comp. Mat., Vol. 12, 1978, 97-115

MI- 29 Kondo, K. and Taki, T.,
Moisture Diffusivity of Unidirectional Composites, J. Comp. Mat., Vol. 16, 1982

MI- 30 Kondo, K. and Taki, T.,
Transverse Moisture Diffusivity of Unidirectionally Fiber-Reinforced Composites, Proc. Japan-US Conf. Comp. Mat., Tokyo, 1981, 308-317

MI- 31 Larder, R. A.,
Application of Three-Dimensional Finite Element Analyses to the Micromechanics of Fibrous Composite Materials, Lawrence Livermore Lab., 7410001-43, Oct. 1974

MI- 32 Laschet, G. and Marechal, E.,
Finite Element Analysis of Interlaminar Stresses in Laminated Composites, 4th World Cong. FEM, Interlaken, Sept. 1984

MI- 33 Laschet, G. and Marechal, E.,
Description d'elements multicouches evaluant les tractions de surface inter laminaires, LTAS Rep. SF-109, Liege, Dec. 1982

MI- 34 Laschet, G. and et al.,
Etude des possibilities de calcul par elements finis des contraintes interlaminaires dans les composites stratifies, LTAS Rep. SF-102, Liege, Dec. 1980

MI- 35 Levy, A. and et al.,
Elastic and Plastic Interlaminar Shear Deformation in Laminated Composites Under Generalized Plane Stress, 3rd Conf. Matrix Meth. Str. Mech., Dayton, 1971

MI- 36 Lien, S. Y.,
Application of the Finite Element Method to Determine the Moisture Content of Composites Under Transient Conditions, 4th Int. Conf. Comp. Mat., Tokyo, Oct. 1982, 971-978

MI- 37 Lumban Tobing, F. E. and et al.,
Finite Element Analysis of Moisture Effects in Graphite-Epoxy Composites, Computers & Structures, Vol. 16, 1-4, 1983, 457-469

MI- 38 Mahishi, J. M. and Adams, D. F.,
Micromechanical Predictions of Crack Initiation Propagation and Crack Growth Resistance in Boron/Aluminium Composites, J. Comp. Mat., Vol. 16, 1982, 457-469

MI- 39 Malcolm, D. J.
Orthogonal Fibre Composites as Micromorphic Materials, Int. J. Engng. Sci., Vol. 20, No. 10, 1982, 1111-24

MI- 40 Mandel, J. A. and et al.,
Micromechanical Studies of Crack Growth in Fiber Reinforced Materials, Engng. Fract. Mech., Vol. 16, No. 5, 1982, 741-754

MI- 41 Miller, A. K. and Adams, D. F.,
Inelastic Finite Element Analysis of a Heterogeneous Medium Exhibiting Temperature and Moisture Dependent Material Properties, Fibre Sci. Technol., Vol. 13, No. 2, 1980, 135-153

MI- 42 Miller, A. K. and Adams, D. E.,
Micromechanical Aspects of the Environmental Behaviour of Composite Materials, Univ. of Wyoming, Rep. UWME-DR-7011111, 1977

MI- 43 Murray, K. H.,
Finite Element Analysis of a Composite Material Interface, NASA-CR-132341, 1973

MI- 44 O'Brien, T. K. and Raju, I. S.,
Strain-Energy-Release Rate Analysis of Delamination Around an Open Hole in
Composite Laminates, 25th Str., Str. Dyn. Mater. Conf., Palm Spring, May 1984
MI- 45 Ouyang, C. and Lu, M.,
On a Micromechanical Model for Cracked Reinforced Composites, *Appl. Math. Mech.*
Vol. *2*, No. 3, 1981, 297-304
MI- 46 Raju, I. S. and Crews, J. H.,
Interlaminar Stress Singularities at a Straight Free Edge in Composite
Laminates, *Computers & Structures*, Vol. *14*, 1-2, 1981, 21-28
MI- 47 Rybicki, E. F.,
An Energy Release Rate Approach for Stable Crack Growth in the Free-Edge
Delamination Problem, *J. Comp. Mat.*, Vol. *11*, 1977, 470-487
MI- 48 Schaffer, B. G. and Adams, D. F.,
Nonlinear Viscoelastic Behaviour of a Composite Material Using a Finite-Element
Micromechanics Analysis, Univ. of Wyoming, UWME-DR-001-101-1, Laramie, June
1980
MI- 49 Stalnaker, D. O. and et al.,
Interlaminar Shear Strain in a Two-Ply Balanced Cord-Rubber Composite, *Exper.
Mech.*, Vol. *20*, 1980, 87-94
MI- 50 Stanton, E. L. and Crain, L. M.,
Interlaminar Stress Gradients and Impact Damage, Proc. Fiber. Comp. Str. Des.,
Plenum Press, 1980, 423-439
MI- 51 Wang, S. S.,
Delamination Fracture from Surface Notch in 45Deg-/0Deg/90Deg Graphite/Epoxy
Composites, 2nd Int. Conf. Comp. Mat., TMS-AIME, 1978, 277-291
MI- 52 Wang, S. S.,
Delamination Crack Growth in Unidirectional Fiber-Reinforced Composites under
Static and Cyclic Loading, Comp. Mat: Test Design, ASTM, STP 674, 1979, 642-663
MI- 53 Wang, S. S.,
An Analysis of Delamination in Angle-Ply Fiber-Reinforced Composites, J. Appl.
Mech., ASME, Vol. *47*, March 1980, 64-70
MI- 54 Whitcomb, J. D.,
Finite Element Analysis of Instability Related Delamination Growth, *J. Comp.
Mat.*, Vol. *15*, 1981, 403-426
MI- 55 Whitcomb, J. D. and Raju, I. S.,
Analysis of Interlaminar Stress in Thick Composite Laminates With and Without
Edge Delamination, NASA TM-85738, 1984
MI- 56 Woolstencroft, D. H. and Curtis, A. R.,
Interface Stresses in Unidirectional Carbon Fibre Composites, Interf. Comp.
Mat., Liverpool Univ., April 1981

Buckling, Post-Buckling (BU)

BU- 1 Agarwal, B. L.,
Post-Buckling Behaviour of Composite Shear Webs, *AIAA J.*, 1980
BU- 2 Agarwal, B. L.,
Post-Buckling Behaviour of Composite Stiffened Curved Panels Loaded in
Compression, 4th SESA Int. Cong. Exp. Mech., Boston, 1980
BU- 3 Chamis, C. C. and Williams, G. C.,
Interply Layer Degradation Effects on Composite Structural Response, NASA Tech.
Mem. 83702, 1984
BU- 4 Cohen, G. A.,
Effect of Transverse Shear Deformation on Anisotropic Plate Buckling, *J. Comp.
Mat.*, Vol. *16*, No. 4, 1982, 301
BU- 5 Crouzet-Pascal, J.,
Buckling Analysis of Laminated Composite Plates, *J. Fibre Sci. Tech.*, Vol. *11*,
No. 6, 1978, 413-446
BU- 6 Davis, R. C.,
Stress Analysis and Buckling of J-Stiffened Graphite-Epoxy Panel, NASA Rep.
TP-1607, 1980
BU- 7 Ishakian, V. G. and Hollaway, L.,

The Stability Analysis of Continuum/Skeletal Fibre/Matrix Composites, *Composites*, Vol. *12*, 1981, 57-64
BU- 8 Janisse, T. C. and Palazotto, A. N.,
Collapse Analysis of Cylindrical Composite Panels With Cutouts, 24th Str., Str. Dyn. Mat. Conf., Lake Tahoe, May 1983, 299-309
BU- 9 Noor, A. K. and Peters, J. M.,
Multiple-Parameter Reduced Basis Technique for Bifurcation and Post-Buckling Analyses of Composite Plates, *Int. J. Num. Meth. Engng.*, Vol. *19*, 1983, 1783-803
BU- 10 Noor, A. K. and Peters, J. M.,
Bifurcation and Post-Buckling Analysis of Laminated Composite Plates via Reduced Basis Technique, *Comp. Meth. Appl. Mech. Engng.*, Vol. *29*, 1981, 271-295
BU- 11 Noor, A. K. and et al.,
Exploiting Symmetries for Efficient Post-Buckling Analysis of Composite Plates, *AIAA J.*, Vol. *15*, No. 1, 1976, 24-32
BU- 12 Raju, K. K. and Rao, G. V.,
Finite Element Analysis of Post-Buckling Behaviour of Cylindrically Orthotropic Circular Plates, *Fibre Sci. Tech.*, Vol. *19*, 1983, 145-154
BU- 13 Raju, K. K. and Rao, G. V.,
Post-Buckling of Cylindrically Orthotropic Circular Plates on Elastic Foundation with Edges Elastically Restrained Against Rotation, *Computers & Structures*, Vol. *18*, No. 6, 1984, 1183-87
BU- 14 Sharifi, P.,
Nonlinear Buckling Analysis of Composite Shells, *AIAA J.*, Vol. *13*, No. 6, 1975
BU- 15 Starnes, J. H. and Haftka, R. T.,
Preliminary Design of Composite Wings for Buckling - Strength and Displacement Constraints, *J. Aircraft*, 1979

Dynamic Analysis (DY)

DY- 1 Aboudi, J.,
The Dynamic Stresses Induced by Moving Interfacial Cracks, *Comp. Meth. Appl. Mech. Engng.*, Vol. *10*, 1977, 303-323
DY- 2 Amijima, S. and et al.,
Stress Wave Propagation in Laminated Composites Having Visco-Elastic Adhesive Layers, ICCM-II, 1978, 1211-33
DY- 3 Bert, C. W. and et al.,
Vibration of Thick Rectangular Plates of Bimodulus Composite Material, J. Appl. Mech., ASME, Vol. *48*, 1981, 371-376
DY- 4 Brockman, R. A.,
Finite Element Analysis of Forced Vibration in Viscoelastic Layered Panel Structures, Univ. of Dayton, UDR-TR-78-31, March 1978
DY- 5 Cawley, P. and Adams, R. D.,
The Predicted and Experimental Natural Modes of Free/Free CFRP Plates, *J. Comp. Mat.*, Vol. *12*, 1978, 336-347
DY- 6 Chamis, C. C. and Lynch, J. E.,
High-Tip-Speed Fiber Composite Fan Blades: Vibration and Stress Analysis, NASA Rep. TM X-71589, 1974
DY- 7 Chen, W. H. and Wu, C. W.,
On Elastodynamic Pfracture Mechanics Analysis of Bi-Material Structures Using Finite Element Method, *Engng. Fracture Mech.*, Vol. *15*, 1981, 155-168
DY- 8 Chung, T. J. and Eidson, R. L.,
Static and Dynamic Analysis of Viscoelastic Fiber-Reinforced Composite Shells in Missile Structures, Alabama Univ., Dept. of Mechanical Eng., 1973
DY- 9 Crawley, E. F.,
The Natural Modes of Graphite/Epoxy Cantilever Plates and Shells, *J. Comp. Mat.*, Vol. *13*, 1979, 195-205
DY- 10 Dharmarajan, S. and Penzes, L. E.,
Dynamic Analysis of Composite Structures, *Composites*, Vol. *8*, No. 1, 1977, 27-32
DY- 11 Frye, M. J. and et al.
Free Vibrations of Layered Spheres, *Acustica*, Vol. *52*, No. 1, 1982, 1-9

DY- 12 Gołub, G. H. and et al.,
Waves in Periodically Structured Media, *J. Comp. Phys.*, Vol. *17*, 1975, 349-357
DY- 13 Jensen, D. W. and et al.,
Vibration of Cantilevered Graphite/Epoxy Plates With Bending-Torsion Coupling,
J. Reinf. Plast. Comp., Vol. *1*, 1982, 254-269
DY- 14 Lakshmninarayana, H. and Sridharamurthy, S.,
Vibration of Laminated Composite Shells of Revolution - A Finite Element
Analysis, NAL-TM-ST-405/198-78, Bangalore, India, Nov. 1978
DY- 15 Laura, P. A. A. and et al.,
Transverse Vibrations of Axisymmetric Polar Orthotropic Circular Plates
Elastically Restrained Against Rotation Along the Edges, *Fibre Sci. Tech.*,
Vol. *15*, No. 1, 1981, 65-77
DY- 16 Mau, S. T. and Pian, T. H. H.,
Linear Dynamic Analyses of Laminated Plates and Shells by the Hybrid-Stress
Finite Element Method, ASRL TR 172-2, MIT, 1973
DY- 17 Mau, S. T. and Witmer, E. A.,
Static, Vibration, and Thermal Stress Analysis of Laminated Plates and Shells
by the Hybrid-Stress Finite-Element Method, ASRL TR 169-2, MIT, Oct. 1972
DY- 18 Mei, C. and Wentz, K. R.,
Large Amplitudes Random Response of Angle-Ply Laminated Composite Plates, 22nd
Str., Str. Dynam. Mater. Conf., Atlanta, April 1981, 559-573
DY- 19 Mei, C. and et al.,
Large-Amplitude Vibration of Laminated Composite Plates of Arbitrary Shape,
24th Str., Str. Dyn. Mater. Conf., Lake Tahoe, May 1983, 685-692
DY- 20 Minagawa, S. and et al.,
Finite Element Analysis of Harmonic Waves in Layered and Fibre-Reinforced
Composites, *Int. J. Num. Meth. Engng.*, Vol. *17*, 1981, 1335-53
DY- 21 Minich, M. D. and Chamis, C. C.,
Analytical Displacements and Vibrations of Cantilevered Unsymmetrical Fiber
Composite Laminates, AIAA Paper 75-757, 1975
DY- 22 Murthy, P. L. N. and Chamis, C. C.,
Dynamic Stress Analysis of Smooth and Notched Fiber Composite Flexural
Specimens, NASA Tech. Mem. 83694, 1984
DY- 23 Nelson, R. B. and Navi, P.,
Harmonic Wave Propagation in Composite Materials, *J. Acoust. Soc. Amer.*, Vol.
57, 1975, 773-781
DY- 24 Nmat-Nasser, S.,
Estimate of Dynamic Properties of Composites by Mixed Finite Elements Method,
in Hybrid Mixed FEM, S. N. Atluri, et al., Wiley, 1983, 243-251
DY- 25 O'Callahan, J. and McElman, J. A.,
Dynamic Composite Laminate Finite Element Analysis, Lowell Univ., AD-A100 643/6,
Mass. March 1981
DY- 26 Pillasch, D. W. and Zak, A. R.,
Finite Element Analysis of a Dynamically Loaded Flat Laminated Plate, Univ. of
Illinois, Rep. AAE 79-4, 1979
DY- 27 Pillasch, D. W. and et al.,
Dynmaic Finite Element Model for Laminated Structures, *Computers & Structures*,
Vol. *16*, 1-4, 1983, 449-455
DY- 28 Raju, K. K. and Rao, V. G.,
Large Amplitude Vibrations of Cylindrically Orthotropic Tapered Circular Plates
Elastically Restrained Against Rotation at the Edges, *Fibre Sci. Tech.*, Vol. *12*,
No. 5, 1979, 395-401
DY- 29 Rand, R. A. and Shen, C. N.,
Optimum Design of Composite Shells Subject to Natural Frequency Constraints,
Computers & Structures, Vol. *3*, No. 2, 1973, 247-263
DY- 30 Reddy, J. N.,
On the Solutions to Forced Motions of Rectangular Composite Plates, *J. Appl.
Mech.*, ASME, Vol. *49*, No. 2, 1982, 403-408
DY- 31 Reddy, J. N.,
Nonlinear Vibration of Layered Composite Plates Including Transverse Shear and
Rotary Inertia, 1981 ASME Vibration Conf., Hartford, Sept. 1981
DY- 32 Reddy, J. N.,

Free Vibration of Antisymmetric, Angle-Ply Laminated Plates, Including
Transverse Shear Deformation by the Finite Element Method, *J. Sound Vib.*, Vol.
66, No. 4, 1979, 565-576

DY- 33 Reddy, J. N.,
Geometrically Nonlinear Transient Analysis of Laminated Composite Plates,
AIAA J., Vol. *21*, No. 4, 1983, 621-629

DY- 34 Reddy, J. N.,
Dynamic Transient Analysis of Layered Anisotropic Composite-Material Plates,
Int. J. Num. Meth. Engng., Vol. *19*, 1983, 237-255

DY- 35 Reddy, J. N.,
Nonlinear Transient Response of Layered Composite Plates, Int. Conf. FEM,
Shanghai, Aug. 1982

DY- 36 Reddy, J. N.,
Transient Response of Laminated, Bimodular-Material, Composite Rectangular
Plates, *J. Com. Mat.*, Vol. *16*, 1982, 139-153

DY- 37 Reddy, J. N. and Chao, W. C.,
Large-Deflection and Large Amplitude Free Vibrations of Laminated Composite
Material Plates, *Computers & Structures*, Vol. *16*, 1981, 341-347

DY- 38 Shivakumar, K. N. and Krishna Murty, A. V.,
A High Precision Ring Element for Vibrations of Laminated Shells, *J. Sound Vib.*,
Vol. *58*, 1978, 311-318

DY- 39 Spilker, R. L. and et al.,
Use of the Hybrid Stress Finite Element Model for the Static and Dynamic
Analysis of Multilayer Composite Plates and Shells, Army Mat. Mech. Res.
Center, ASRL-TR-181-2, Watertown, Sept. 1976

DY- 40 Sun, C. T. and Whitney, J. M.,
Theories for the Dynamic Response of Laminated Plates, *AIAA J.*, Vol. *11*, 1973,
178

DY- 41 Tabaddor, F. and Stafford, J. R.,
Nonlinear Vibration of Cord-Reinforced Composite Shells, *Computers & Structures*,
Vol. *13*, 1981, 737-743

DY- 42 Ten, K. K. and Huang, C. C.,
The Vibrations of Generally Orthotropic Beams: A Finite Element Approach,
J. Sound Vib., Vol. *62*, No. 2, 1979, 195-206

DY- 43 Thornton, E. A. and Clary, R. R.,
A Correlation Study of Finite Element Modelling for Vibration of Composite
Material Panels, in Comp. Mat., Testing, Design, ASTM, STP 546, 1974, 111-128

DY- 44 Wang, S. S.,
Delamination Crack Growth in Unidirectional Fiber-Reinforced Composites under
Static and Cyclic Loading, Comp. Mat.: Test Design, ASTM, STP 674, 1979, 642-663

DY- 45 Wen-Hwa, C. and Chei-Wei, W.,
On Elastodynamic Fracture Mechanics Analysis of Bi-Material Structures Using
Finite Element Method, Engng. Fract. Mech., Vol. *15*, 1-2, 1981, 155-168

DY- 46 Winter, R. and et al.,
Crash Simulation of Composite and Aluminium Helicopter Fuselages Using a
Finite Element Program, 20th Str., Str. Dyn. Mater. Conf., St. Louis, 1979

DY- 47 Yamada, Y. and Okumura, H.,
Finite Element Analysis of Stress and Strain Singularity Eigenstate in
Inhomogeneous Media or Composite Materials, in Hybrid Mixed FFM, S. N. Atluri,
et al., Wiley, 1983, 325-343

DY- 48 Yang, W. H. and Lee, E. H.,
Modal Analysis of Floquet Waves in Composite Materials, *J. Appl. Mech.*, Vol. *41*,
1974, 429-433

DY- 49 Zak, A. R. and Pillasch, D. W.,
Finite Element Analysis of a Dynamically Loaded Laminated Plate, ARBRL-CR-00433,
July 1980

Element Library (EL)

EL- 1 Apostal, M. C.,
Development of an Anisotropic Singularity Finite Element Utilizing the Hybrid

Displacement Method, Ph D Dissert., State Univ. of New York, Buffalo, 1974
EL- 2 Apostal, M. C. and et al.,
Anisotropic Hybrid Displacement Singularity Element, J. Struct. Div., ASCE,
Feb. 1977, 335-354
EL- 3 Avramidis, I. E.,
A Two-Component Finite Element for Determining Transverse Loads in
Unidirectional Composites, Ingenieur-Archiv., Vol. 48, No. 1, 1979, 13-26
EL- 4 Bartelds, G.,
Sandwich and Composite Elements, World Cong. FEM Str. Mech., Bournemouth, 1975
EL- 5 Baumann, E.,
An Evaluation of ASKA Anisotropic Finite Elements as Utilized for Analysis of
Structures Composed of Advanced Composite Materials, 4th Int. Sem. Comp.
Aspects FEM, Aug. 1977
EL- 6 Brunner, M.,
QUAD4C- A Failure Prediction Element for Fibre Reinforced Composite in NASTRAN,
NASTRAN User's Conf., Munich, May 1979
EL- 7 Carnoy, E. and Guennoun, N.,
Elements finis de coque et de volume multicouche developpes en series de
Fouries pour l'analyse de stabilite, LTAS Rep. SF-110, Liege, Dec. 1982
EL- 8 Epstein, M. and Huttelmaier, H. P.,
A Finite Element Formulation for Multilayered and Thick Plates, *Computers &
Structures*, Vol. *16*, No. 5, 1983, 645-650
EL- 9 Farley, G. L. and Baker, D. J.,
Graphics and Composite Material Computer Program Enhancements for Use for SPAR,
NASA Tech. Mem., TM-80208, 1980
EL- 10 Foschi, R. O. and Barrett, J. D.,
Stress Intensity Factors in Anisotropic Plates Using Singular Isoparametric
Elements, *Int. J. Num. Meth. Engng.*, Vol. *10*, 1976, 1281-87
EL- 11 Giavotto, V. and et al.,
Anisotropic Beam Theory and Applications, *Computers & Structures*, Vol. *16*,
1-4, 1983, 403-413
EL- 12 Harris, A. and et al.,
A Multilayer, Traction-Free-Edge, Quadrilateral Warping Element for Stress
Analysis of Composite Plates and Shells, Rep. ASRL-TR-193-1, MIT, 193-1, 1979
EL- 13 Heppler, G. and Hansen, J. S.,
Mixed Mode Fracture Analysis of Rectilinear Anisotropic Plates by High Order
Finite Elements, *Int. J. Num. Meth. Engng.*, Vol. *17*, 1981, 445-464
EL- 14 Heppler, G. R. and Hansen, J. S.,
A High Precision Singular Finite Element for Fracture Analysis of Composite
Structures, 3rd Int. Conf. Comp. Mat., Pergamon Press, Paris, Aug. 1980
EL-15 Hinton, E.,
the Flexural Analysis of Laminated Composites Using a Parabolic Isometric
(SIC) Plate Bending Element, *Int. J. Num. Meth. Engng.*, Vol. *11*, 1977, 174-179
EL- 16 Holt, J. S. and et al.,
Displacement Oscillation in Plane Quadratic Isoparametric Elements in
Orthotropic Situations, *Int. J. Num. Meth. Engng.*, Vol. *14*, No. 6, 1979, 913-920
EL- 17 Huttelmaier, H. P. and Epstein, M.,
A Nonlinear Multilayered Plate Finite Element, Fenomech, Stuttgart, Sept. 1984
EL- 18 Jones, R. and et al.,
Analysis of Multi-Layer Laminates Using Three-Dimensional Super-Elements, *Int.
J. Num. Meth. Engng.*, Vol. *20*, 1984, 583-587
EL- 19 Kwok, W. L. and Cheung, Y. K.,
Analysis of Circular and Annular Laminated Thick Plates, in Finite Elem. Meth.
Engng., Univ. NSW, Australia, 1974, 177-194
EL- 20 Lakshminarayana, H. and Murthy, S. S.,
A Shear-Flexible Triangular Finite Element Model for Laminated Composite
Plates, *Int. J. Num. Meth. Engng.*, Vol. *20*, 1984, 591-623
EL- 21 Lakshminarayana, H. V. and Viswanath, S.,
A High Precision Triangular Laminated Cylindrical Shell Finite Element,
Computers & Structures, Vol. *8*, 1978, 633-640
EL- 22 Laschet, G.,
Description d'un element multicouche pour l$\frac{3}{8}$analyse des resines armees, LTAS

Rep. SF-98, Liege, Aug. 1980
EL- 23 Laschet, G. and Gilman, H.,
Description d'un element multicouche tridimensional specialise pour l'etude des coques, LTAS Rep. SF-117, Liege, Apr. 1983
EL- 24 Laschet, G. and Marechal, F.,
Description d'elements multicouches evaluant les tractions de surface inter laminaires, LTAS Rep. SF-109, Liege, Dec. 1982
EL- 25 Laschet, G. and et al.,
Element multicouche pour l½analyse des resines armees, LTAS Rep. SF-91, Liege, March 1980
EL- 26 Lin, K. Y. and Tong, P.,
A Hybrid Crack Element for the Fracture Mechanics Analysis of Composite Materials, in *Num. Meth. Fract. Mech.*, Swansea, 1978
EL- 27 Mathers, M. D.,
Isoparametric Shell Finite Elements for Automotive Composite Structures, 3rd Int. Conf. Vehicle Str. Mech. Troy, Oct. 1979, 105-113
EL- 28 Minich, M. D. and Chamis, C. C.,
Doubly-Curved Variable-Thickness Isoparametric Heterogeneous Finite Element, *Computers & Structures*, Vol. *7*, 1977, 295-301
EL- 29 Murthy, S. S. and Gallagher, R. H.,
Anisotropic Cylindrical Shell Element Based on Discrete Kirchhoff Theory, *Int. J. Num. Meth. Engng.*, Vol. *19*, 1983, 1805-23
EL- 30 Nishioka, T. and Atluri, S. N.,
Stress Analysis of Holes in Angle-Ply Laminates: An Efficient Assumed Stress "Special-Hole-Element" Appraoch and a Simple Estimation Method, *Computers & Structures*, Vol. *15*, No. 2, 1982, 135-147
EL- 31 Noor, A. K. and Anderson, C. M.,
Mixed Isoparametric Laminated Composite Shell Elements, World Cong FEM Str. Mech., Bournemouth, Oct. 1975
EL- 32 Nopratvarakorn, V. and Huang, J. C.,
Quadrilateral Plate Element for Laminated Structure, *AIAA J.*, Vol. *17*, 1979, 95
EL- 33 Owen, D. R. J. and Figueiras, J. A.,
Elasto-Plastic Analysis of Anisotropic Plates and Shells by the Semiloof Element, *Int. J. Num. Meth. Engng.*, Vol. *19*, 1983, 521-539
EL- 34 Patil, S. S. and Rao, A. K.,
Special Finite Elements for Flaw and Discontinuity Stresses in Composites, 2nd Int. Conf. Str. Mech. Reactor Tech., Berlin, Sept. 1973
EL- 35 Ramsay, A.,
Evaluation of NASTRAN Elements in the Analysis of Fibre-Reinforced Composite, MSC/NASTRAN Users' Conf., Munich, Apr. 1982
EL- 36 Ranga Rao, M. N. and et al.,
On Applicability of Reduced Bending Stiffness Approximation to Unsymmetric Laminates, Dept. Aeronaut. Engng. Rep. AE 365S, Bangalore, 1982
EL- 37 Rao, K. P.,
A Rectangular Laminated Anisotropic Shallow Thin Shell Finite Element, *Comp. Meth. Appl. Mech. Engng.*, Vol. *15*, 1978, 13-33
EL- 38 Reddy, N. J.,
A Penalty Plate-Bending Element for the Analysis of Laminated Composite Plates, *Int. J. Num. Meth. Engng.*, Vol. *15*, 1980, 1187-206
EL- 39 Shivakumar, K. N. and Krishna Murty, A. V.,
A High Precision Ring Element for Vibrations of Laminated Shells, *J. Sound Vib.*, Vol. *58*, 1978, 311-318
EL- 40 Spilker, R. L.,
Hybrid-Stress Eight-Node Elements for Thin and Thick Multilayer Laminated Plates, *Int. J. Num. Meth. Engng.*, Vol. *18*, 1982, 801-828
EL- 41 Spilker, R. L.,
An Invariant Eight-Node Hybrid-Stress Element for Thin and Thick Multilayer Laminated Plates, *Int. J. Num. Meth. Engng.*, Vol. *20*, 1984, 573-587
EL- 42 Spilker, R. L.,
High Order Three-Dimensional Hybrid-Stress Elements for Thick-Plate Analysis, *Int. J. Num. Meth. Engng.*, Vol. *17*, 1981, 53-69
EL- 43 Spilker, R. L.,

A Traction-Free-Edge Hybrid-Stress Element for the Analysis of Edge Effects in Cross-Ply Laminates, *Computers & Structures*, Vol. *12*, 1980, 167-179
EL- 44 Spilker, R. L. and *et al.*,
Alternate Hybrid-Stress Elements for Analysis of Multilayer Composite Plates, J. Comp. Mat., Vol. *11*, 1977, 51-70
EL- 45 Tong, P.,
A Hybrid Crack Element for Rectilinear Anisotropic Material, Int. J. Num. Meth. Engng., Vol. *11*, No. 2, 1977, 377-403
EL- 46 Venkatesh, A. and Rao, K. P.,
A Laminated Anisotropic Curved Beam and Shell Stiffening Finite Element, *Computers & Structures*, Vol. *15*, No. 2, 1982, 197-201
EL- 47 Venkatesh, A. and Rao, K. P.,
A Doubly Curved Quadrilateral Finite Element for the Analysis of Laminated Anisotropic Thin Shells of Revolution, *Computers & Structures*, Vol. *12*, 1980, 825-832
EL- 48 Venkatesh, A. and Rao, K. P.,
A Curved Laminated Anisotropic Beam and Shell Stiffening Finite Element, Dept. of Aeronaut. Rep. AE 356S, Bangalore, 1980
EL- 49 Venkatesh, A. and Rao, K. P.,
A Quadrilateral Layered Anisotropic Shell of Revolution Finite Element, *Computers & Structures*, Vol. *12*, 1980, 825-832
EL- 50 Venkatesh, A. and Rao, K. P.,
Laminated Anisotropic Curved Beam Finite Elements, Dept. of Aeronaut. Rep. AE 362S, Bangalore, 1981
EL- 51 Wang, S. S. and Stango, R. J.,
Optimally Discretized Finite Elements for Boundary-Layer Stresses in Composite Laminates, *AIAA J.*, Vol. *21*, No. 4, 1983, 614-620
EL- 52 Wang, T. K. and Yuan, U.,
A New Hybrid-Stress Element for Multilayer Laminates, *J. Reinf. Plast. Comp.*, Vol. *1*, 1982, 124-130
EL- 53 Wilt, J. C.,
Use of Composite Materials in NASTRAN, 24th Str., Str. Dyn. Mater. Conf., Lake Tahoe, May 1983, 612-622
EL- 54 Yamada, Y. and Okumura, H.,
Analysis of Local Stress in Composite Materials by the 3D Finite Element, Proc. Japan-US Symp. Comp. Mat., Tokyo, 1981, 55-64
EL- 55 Yuan, Y.,
Two Kinds of Hybrid-Stress Element Analysis of Multilayer Lamintes Including Transverse Shear Effect, Thesis, Beijing Inst. Aeronaut. Astron., China, 1982

Fracture Mechanics Analysis (FM)

FM- 1 Aberson, J. A.
Characterization of Crack Growth in Bonded Structure, 12th Ann. Meet. Soc. Engng. Sci., Univ. Texas, Austin, Oct. 1975
FM- 2 Aberson, J. A.,
Fracture Control for Composite Lamintes, Comp. Mater: The Infl. Mech. Fail Design, Cape Code, Sept. 1976
FM- 3 Aberson, J. A.,
Fracture in Composites, Lockheed-Georgia Co., Rep SMN391, Febr. 1976
FM- 4 Aboudi, J.,
The Dynamic Stresses Induced by Moving Interfacial Cracks, *Comp. Meth. Appl. Mech. Engng.*, Vol. *10*, 1977, 303-323
FM- 5 Adams, D. F.,
A Micromechanical Analysis of Crack Propagation in an Elastoplastic Composite Material, *Fibre Sci. Tech.*, Vol. *7*, No. 4, 1974, 237
FM- 6 Adams, D. F. and Murphy, D. P.,
Analysis of Crack Propagation as an Energy Absorption Mechanism in Metal Matrix Composites, Univ. of Wyoming, DR-101-102-1, 1981
FM- 7 Adams, D. S. and Herakovich, C. T.,
Influence of Damage on the Thermal Response of Graphite-Epoxy Laminates, *J.*

Thermal Stresses, Vol. *7*, 1984, 91-103
FM- 8 Agrawal, S.,
A Technique to Predict the Post Cracking Behaviour of RCC Plated Structures, in Engng. Software, Springer Verlag, Vol. *III*, 1983, 580-591
FM- 9 Agrawal, S.,
Prediction of Post Cracking Behaviour of RCC Plate Structures, Ph D Thesis, Univ. of Roorkee, India, 1979
FM- 10 Anderson, G. P.,
Applied Adhesive Fracture Mechanics, PH D Dissert., Univ. of Utah, Salt Lake City, Utah, 1973
FM- 11 Anderson, G. P. and et al.,
Finite Element in Numerical Analysis of Adhesive Fracture, *Int. J. Fract.*, Vol. *9*, No. 3, 1973, 335-336
FM- 12 Aronsson, C. G. and Backlund, J.,
Damage Analysis of Fibrous Composites, 4th Int. Conf. Comp. Mat., Tokyo, 1982, 607
FM- 13 Aronsson, C. G. and Backlund, J.,
FRACOM - A Computer Program for Fracture Analysis of Composites, Inst. of Technol., Linkoping, IKP-R-167, 1980
FM- 14 Aronsson, C. G.,
Tensile Fracture of Composite Laminates with Holes and Cracks, Thesis, The Royal Inst. of Tech., Stockholm, 84-5, 1984
FM- 15 Aronsson, C. G.,
Stacking Sequence Effects on Fracture of Notched Carbon/Epoxy Composites, Int. Symp. Comp.: Mater. Engng., Univ. Delaware, Delaware, Sept. 1984
FM- 16 Aronsson, C. G. and Backlund, J.,
Damage Mechanics Analysis of Matrix Effects in Notched Laminates, Symp. Comp. Mater.: Fract. Fatigue, ASTM, Ft. Worth, Oct. 1984
FM- 17 Backlund, J.,
Fracture Analysis of Notched Composites, *Computers & Structures*, Vol. *13*, 1981, 145-154
FM- 18 Barker, R. M. and MacLaughlin, T. F.,
Stress Concentrations Near a Discontinuity in Fibrous Composites, *J. Comp. Mat.*, Vol. *5*, 1971, 492-503
FM- 19 Bennett, S. J. and et al.,
Adhesive Fracture Mechanics, Int. J. Fract., Vol. *10*, No. 1, 1974, 33-43
FM- 20 Berg C. A. and Salama, M.,
Fatigue of Prenotched Graphite Fiber Composites in Compression, *Text Res. J.*, Vol. *42*, No. 4, 1972, 222-238
FM- 21 Bigelow, C. A. and et al.,
Fracture of Metal Matrix Composites, 11th Southeast Conf. Theor. Appl. Mech., Huntsville, Apr. 1982
FM- 22 Braun, H. and Herrmann, K.,
Analysis of Thermal Cracking of Unidirectionally Reinforced Composite Structures in the Micromechanical Range, Proc. 5th Int. Conf. Fract., ICF 5, Pergamon Press, 1981, 485-493
FM- 23 Braun, H. and Herrmann, K.,
Quasistatic Crack Growth in Unit Cells of a Self-Stressed Fiber-Reinforced Composite, *ZAMM*, Vol. *60*, No. 6, 1980, 105-106
FM- 24 Braun, H. and Herrmann, K.,
Numerical Calculation of Strain Energy Release Rates for Thermal Cracks in Fibre-Reinforced Materials, 2nd Int. Conf. Num. Meth. Fract. Mech., Swansea, 1980, 207-237
FM- 25 Braun, H. and et al.,
Finite Element Analysis of a Quasistatic Crack Extension in a Unit Cell of a Fibre-Reinforced Material, *Int. J. Fract.*, Vol. *14*, 1978
FM- 26 Brown, G. E.,
Progressive Failure ofAdvanced Composite Laminates Using the Finite Element Method, M. S. Thesis, Univ. of Utah, 1976
FM- 27 Brunner, M.,
QUAD4C - A Failure Prediction Element for Fibre Reinforced Composite in NASTRAN, NASTRAN User's Conf., Munich, May 1979

FM- 28 Buczek, M. B.,
Finite Element Models for Predicting Crack Growth Characteristics in Composite Materials, M. S. Thesis, Eng. & Sci. Mech., VPI 82-29, Blacksburg, Oct. 1982

FM- 29 Buczek, M. B. and Herakovich, C. T.,
Finite Element Models for Predicting Crack Growth Characteristics in Composite Materials, Virginia Polyt. Inst., VPI E-82-29, Blacksburg, 1982

FM- 30 Buczek, M. B. and Herakovich, C. T.,
Direction of Crack Growth in Fibrous Composites, in *Mech. Comp. Mater.*, AMD, Vol. *58*, 1983, 75-82

FM- 31 Chamis, C. C. and Sinclair, J. H.,
Mechanical Behaviour and Fracture Characteristics of Off-Axis Fiber Composites, Part II - Theory and Comparisons, NASA Rep TP-1082, 1978

FM- 32 Chang, F. K. and *et al.*,
Failure of Composite Laminates Containing Pin Loaded Holes - Method of Solution, *J. Comp. Mater.*, May 1984, 255-278

FM- 33 Chang, F. K. and*et al.*,
Failure Strength of Nonlinearly Elastic Laminates Containing Pin-Loaded Holes, *J. Comp. Mater.*, Sept. 1984

FM- 34 Chen, W. H.,
On the J-Integral for Nonhomogeneous Cracked Composites, 5th Int. Conf. Str. Mech. Reactor Tech., Berlin, Aug. 1979, M9/4

FM- 35 Chen, W. H.,
A Hybrid-Displacement Finite Element Analysis for Nonhomogeneous Cracked Composites, 2nd Nat. Conf. Theor. Appl. Mech., ROC, 1978, 122-136

FM- 36 Chen, W. H. and Wu, C. W.,
On Elastodynamic Fracture Mechanics Analysis of Bi-Material Structures Using Finite Element Method, *Engng. Fracture Mech.*, Vol. *15*, 1981, 155-168

FM- 37 Chen, W. H. and Wu, C. W.,
Finite Element Analysis of Bonded Materials With a Crack Perpendicular to the Interface, 3rd Nat. Conf. Theor. Appl. Mech., ROC, 1979, 95-107

FM- 38 Chou, S. C. and *et al.*,
Post-Failure Behaviour of Laminates: II - Stress Concentration, *J. Comp. Mat.*, Vol. *11*, 1977, 71-78

FM- 39 Chu, C. S. and Freyre, O. L.,
Failure Stress Correlation of Composite Laminates Containing a Crack, *J. Aircraft*, Vol. *19*, No. 2, 1982, 164-168

FM- 40 Crossman, F. W. and Wang, A. S. D.,
The Dependence of Transverse Cracking and Delamination on Ply Thickness in Graphite-Epoxy Laminates, ASTM-STP 775, 1982, 118

FM- 41 Crossman, F. W. and *et al.*
Initiation and Growth of Transverse Cracks and Edge Delamination in Composite Laminates, Part II Experiment Correlation, J. Comp. Mater., Suppl. Vol., 1980

FM- 42 Crossman, F. W. and *et al.*,
Influence of Ply Thickness on Damage Accumulation and Final Failure, in Adv. Aerosp. Str. Mater. Dynam., ASME, AD-06, 1983, 215

FM- 43 Dharan, C. K. H.,
Fracture Mechanics of Composite Materials, *J. Engng. Mat. Tech.*, Vol. *100*, 1978, 23

FM- 44 Dickson, J. N. and *et al.*,
Development of an Understanding of the Fatigue Phenomena of Bonded and Bolted Joints in Advanced Filamentary Composite Materials, WPAFB, Rep AFFDL-TR-72-64, Ohio, 1972

FM- 45 Dvorak, G. J. and Bahei-El-Din, Y. A.,
Plastic Yielding at a Crack Tip in a Laminated B-Al Plate, 16th Ann. Meeting. Soc. Engng. Sci., Northw. Univ., Evanston, Sept. 1979

FM- 46 Foschi, R. O. and Barrett, J. D.,
Stress Intensity Factors in Anisotropic Plates Using Singular Isoparametric Elements, *Int. J. Num. Meth. Engng.*, Vol. *10*, 1976, 1281-87

FM- 47 Gallo, R. L. and Palazotto, A. N.,
Approximate Design Analysis of Failure Within Composite Laminated Plates, J. Mech. Design, ASME, Vol. *104*, No. 3, 1982, 604-611

FM- 48 Gradin, P. A.,
A Fracture Criterion for Edge-Bonded Bimaterial Bodies, *J. Comp. Mat.*, Vol. *16*, 1982, 448-456

FM- 49 Guess, T. R. and Gerstle, F. P.,
Deformation and Fracture of Resin Matrix Composites in Combined Stress States, *J. Comp. Mat.*, Vol. *11*, 1977, 146-163

FM- 50 Hayashi, I. and Iwamoto, J.,
Stress Distribution Near a Broken Point of Filament in Unidirectionally Fiber Reinforced Composites, 17th Japan Cong. Mat. Res., Tokyo, 1973, 162-166

FM- 51 Hellan, K. and Storkersen, B.,
On a Multiparametric Characterization of Cracking in Fibrous Composites, *Int. J. Fract.*, Vol. *23*, No. 4, 1983, R161-67

FM- 52 Henze, E. and Roth, S.,
Practical Finite Element Method of Failure Prediction for Composite Material Structures, Proc. AGARD Conf., Munich, Germany, Oct. 1974

FM- 53 Heppler, G. and Hansen, J. S.,
Mixed Mode Fracture Analysis of Rectilinear Anisotropic Plates by High Order Finite Elements, *Int. J. Num. Meth. Engng.*, Vol. *17*, 1981, 445-464

FM- 54 Heppler, G. R. and Hansen, J. S.,
High Accuracy Finite Element of Cracked Rectilinear Anisotropic Structures Subject to Biaxial Stress, 2nd Conf. Num. Meth. Fract. Mech., Swansea, July 1980, 223-237

FM- 55 Heppler, G. R. and Hansen, J. S.,
A High Precision Singular Finite Element for Fracture Analysis of Composite Structures, 3rd Int. Conf. Comp. Mat., Pergamon Press, Paris, Aug. 1980

FM- 56 Heppler, G. R. and *et al.*,
Stress Intensity Factor Calculation for Designing With Fiber Reinforced Composite Materials, 24th Str., Str. Dynam. Mat. Conf., Lake Tahoe, 1983, 161-170

FM- 57 Herakovich, C. T. and Bergner, H. W.,
Finite Element Stress Analysis of a Notched Coupon Specimen for In-Plane Shear Behaviour of Composites, *Composites*, Vol. *11*, No. 3, 1980, 149-154

FM- 58 Herakovich, C. T. and *et al.*,
Failure Analysis of Composite Laminates with Free Edges, in Modern Dev. Comp. Mat. Struct., ASME, 1979, 53-66

FM- 59 Herrman, K. and *et al.*,
Finite Element Analysis and Experimental Verification of Quasistatic Thermal Crack Growth in a Two-Phase Medium, *Anal. Exper. Fract. Mech.*, Rome, June 1980

FM- 60 Herrmann, K.,
Analysis of a Slow Thermal Crack Propagation in a Composite Structure by Consideration of Strain Hardening and Softening in the Plastic Zones, EUROMECH 91, Warschau, 1977

FM- 61 Herrmann, K.,
Interaction of Cracks and Self-Stresses in a Composite Structure, 2nd Int. Symp. Contin. Mod. Discr. Systm., SM Study 12, 1978, 313-338

FM- 62 Herrmann, K.,
Quasistatic Thermal Crack Growth in the Viscoelastic Matrix Material of a Brittle Fiber Reinforced Unit Cell, *Mech. Res. Comm.*, Vol. *8*, 1981, 97-104

FM- 63 Herrmann, K.,
Curved Thermal Crack Growth in the Interfaces of a Unidirectional Carbon-Aluminium Composite, in *Adv. Mech. Comp. Mater.*, Pergamon Press, New York, 1983, 383-397

FM- 64 Herrmann, K. and Braun, H.,
Quasistatic Thermal Crack Growth in Unidirectionally Fiber Reinforced Composite Materials, *Emgng. Fract. Mech.*, Vol. *18*, No. 5, 1983, 975-996

FM- 65 Herrmann, J. and Braun, H.
Analysis of Cracks in Composite Structures Subjected by Thermal Loading, 1st USA-USSR Symp. Fract. Comp. Mat., Sijthoff, 1979, 171-192

FM- 66 Herrmann, K. and Grebner, H.,
Analysis and Experiment of Curved Thermal Crack Growth in Different Shaped Dissimilar Media, Conf. Appl. Fract. Mech. Mater. Str., Freiburg, June 1983

FM- 67 Herrmann, K. and Grebner, H.,
Thermal Crack Propagation in Different Shaped Two-Phase Composite Structures:
Analysis and Experiment, Int. Symp. Fract. Mech., Beijing, Peking, China,
Nov. 1983, 1063-68

FM- 68 Herrmann, K. and Grebner, H.,
Slow Thermal Crack Growth in Thermally Loaded Two-Phase Composite Structures
Containing Inner Stress Concentrators, 5th Europ. Conf. Fract. (ECF5), Lissabon,
Portugal, Sept. 1984

FM- 69 Herrmann, K. and Mihovsky, I.,
Plastic Behaviour of Fibre-Reinforced Composites and Fracture Effects, 4th Nat.
Cong. Theor. Appl. Mech., Varna, Bulgaria, Sept. 1981, 431-436

FM- 70 Herrmann, K. and Mihovsky, I. M.,
Plastic Behaviour of Fibre-Reinforced Composites and Fracture Effects, *Engng.
Trans.*, Vol. *31*, 1983, 165-177

FM- 71 Herrmann, K. and Mihovsky, I. M.,
Elastic-Plastic Interaction of Dugdale Type Cracks and Plastified Matrix
Material in Self-Stressed Fibre-Reinforced Composites, *ZAMM*, Vol. *63*, 1983,
T167-69

FM- 72 Herrmann, K. and Strathmeier, U.,
Quasistatic Extension of an Interface Crack, 4th Europ. Conf. Fract., ECF4,
Leoben, Austria, Sept. 1982

FM- 73 Herrmann, K. and *et al.*,
Comparison of Experimental and Numerical Investigations Concerning Thermal
Cracking of Dissimilar Materials, *Int. J. Fract.*, Vol. *15*, 1979, 187-190

FM- 74 Herrmann, K. and *et al.*,
Finite Element Analysis and Experimental Verification of Quasistatic Thermal
Crack Growth in a Two-Phase Medium, Int. Conf. Anal. Exper. Fract. Mech., Rome,
June 1980

FM- 75 Hilton, P. D. and Sih, G. C.,
Application of the Finite Element Method to the Calculations of Stress Intensity
Factors, Meth. Anal. Solut. Crack Pr., Noordhoff, Leyden, 1973, 426-483

FM- 76 Humphreys, E. A. and Rosen, B. W.,
Development of a Realistic Stress Analysis for Fatigue Analysis of Notched
Composite Laminates, NASA, CR-159119, May 1979

FM- 77 Johnson, W. S. and *et al.*,
Experimental and Analytical Investigation of the Fracture Process of Boron/
Aluminum Laminates Containing Notches, NASA Tech. Paper 2187, Sept. 1983

FM- 78 Jones, R. and Callinan, R. J.,
Analysis of Compression Failures in Fibre Composite Laminates, 4th Int. Conf.
Comp. Mat., Tokyo, Oct. 1982, 447-453

FM- 79 Karlak, R. F. and Crossman, F. W.,
Failure Mechanisms in Composite Systems, Palo Alto Res. Lab. LMSC-D457462,
Palo Alto, Aug. 1975

FM- 80 Karlak, R. F. and Crossman, F. W.,
Interface Failures in Composites, 6th AIME Annual Spring Meet., Pittsburgh,
1974, 119-130

FM- 81 Ko, W. L.,
Finite Element Microscopic Stress Analysis of Cracked Composite Systems, *J.
Comp. Mat.*, Vol. *12*, 1978, 97-115

FM- 82 Lakshminarayana, H.,
A Symmetric Rail Shear Test for Mode II Fracture Toughness (GIIC) of Composite
Materials - Finite Element Analysis, *J. Comp. Mater.*, Vol. *18*, May 1984, 227

FM- 83 Lakshminarayana, H. and *et al.*,
On the Behaviour of Cracks in Laminated Composite Panels, in Fract. Mech.
Engng. Appl., Sijthoff, G. C. Sih, ed., 1979

FM- 84 Lakshminarayana, H. V.,
Stress Distribution Around a Semi-Circular Edge-Notch in a Finite Size
Laminated Composite Plate Under Uniaxial Tension, *J. Comp. Mat.*, Vol. *17*, No.
4, 1983, 357-367

FM- 85 Lakshminarayana, H. V. and *et al.*,
On a Finite Element Model for the Analysis of Through Cracks in Laminated
Anisotropic Cylindrical Shells, *Engng. Fract. Mech.*, Vol. *14*, No. 4, 1981,

697-712
FM- 86 Lau, K. J. and Chow, C. L.,
A Finite Element Method of Stress Intensity Factor Determination in Cracked Orthotropic Plates, Int. Conf. Fract. Mech. Tech., Sijthoff Publ., Hong Kong, 1977
FM- 87 Lee, J. D.,
Three Dimensional Finite Element Analysis of Damage Accumulation in Composite Laminate, Computers & Structures, Vol. 15, No. 3, 1982, 335-350
FM- 88 Leverenz, R. K.,
A Finite Element Stress Analysis of a Crack in a Bi-Material Plate, Int. J. Fract., Vol. 8, No. 3, 1972, 311-324
FM- 89 Lin, K. Y.,
The Stress Intensity of a Crack at an Interface Between Two Materials, ASRL TR162-5, Aeroelast. Str. Res. Lab., MIT, June 1973
FM- 90 Lin, K. Y. and Mar. J. W.,
Finite Element Analysis of Stress Intensity Factor for Cracks at Bi-Material Interface, Int. J. Fract., Vol. 12, 1976, 521-531
FM- 91 Lin, K. Y. and Tong, P.,
A Hybrid Crack Element for the Fracture Mechanics Analysis of Composite Materials, in Num. Meth. Fract. Mech., Swansea, 1978
FM- 92 Mahishi, J. M. and Adams, D. F.,
Micromechanical Predictions of Crack Initiation Propagation and Crack Growth Resistance in Boron/Aluminum Composites, J. Comp. Mat., Vol. 16, 1982, 457-469
FM- 93 Mandel, J. A. and et al.,
Micromechanical Studies of Crack Growth in Fiber Reinforced Materials, Engng. Fract. Mech., Vol. 16, No. 5, 1982, 741-754
FM- 94 McLay, R. W.,
Analaysis of Fracture in a Linear Composite, 1st Int. Conf. Num. Meth. Fract. Mech., Swansea, Jan. 1978, 721-732
FM- 95 McLay, R. W. and Murphy, M. C.,
An Energy Search Technique for Composite Fracture, in Interdiscip. Finite Elem. Anal., Cornell Univ., 1981, 119-135
FM- 96 Mitchell, R. A. and et al.,
Analysis of Composite Reinforced Cutouts and Cracks, 15th Str., Str. Dyn. Mater. Conf., Las Vegas, 1974, 74-377
FM- 97 Mitchell, R. A. and et al.,
Composite-Overlay Reinforcement of Cutouts and Cracks in Metal Sheet, Engng. Mech. Sec., NBSIR-73-201, NBS, Washington, D.C., 1973
FM- 98 Miyoshi, T. and et al.,
Evaluation of Fracture Toughness of Composite Materials, 5th Nat. Symp. Finite Elem., Tokyo, June 1977
FM- 99 Nagarkar, A. P. and Herakovich, C. T.,
Nonlinear Temperature Dependent Failure Analysis of Finite Width Composite Laminates, NASA CR-162868, Dec. 1979
FM- 100 Nishioka, T. and Atluri, S. N.,
Multilayer-Stress-Hybrid-Finite Element Method for Fracture Analysis of Angle-Ply Laminates, Proc. 2nd Conf. Num. Meth. Fract. Mech., Swansea, 1980, 195-205
FM- 101 Nishioka, T. and Atluri, S. N.,
A Simple Estimation Method of Stress Intensity Factors for Through-Cracks in Angle-Ply Laminates, Engng. Fract. Mech., Vol. 16, No. 4, 1982, 573-583
FM- 102 Nishioka, T. and Atluri, S. N.,
Assumed Stress Finite Element Analysis of Through-Cracks in Angle-Ply Laminates, AIAA J., Vol. 18, No. 9, 1980, 1125-32
FM- 103 Nishioka, T. and Atluri, S. N.,
Fracture-Stress Analysis of Through-Cracks in Angle-Ply Laminates: An Efficient Assumed-Stress Finite Element Approach, 20th Str., Str. Dyn. Mater. Conf., St. Louis, Apr. 1979
FM- 104 Nishioka, T. and Atluri, S. N.,
Fracture Analyses of Angle Ply Laminates, Proc. ICF5, Adv. Fract. Res., Pergamon Press, Vol. 3, 1981, 1245-52
FM- 105 Nishioka, T. and Atluri, S. N.,

Fracture-Stress Analysis of Holes and Cracks in Angle-Ply Laminates: An
Efficient Assumed Stress Finite Element Approach - Part III, 21st Str., Str.
Dyn. Mater. Conf., Seattle, May 1980
FM- 106 Nuismer, R. J. and Brown, G. E.,
Progressive Failure of Notched Composite Laminates Using Finite Elements, in
Adv. Engng. Sci., NASA CP-2001, Vol. 1, 1976
FM- 107 Nuismer, R. J. and Labor, J. D.,
Applications of the Average Stress Failure Criterion: Part I - Tension, *J. Comp. Mat.*, Vol. 12, 1978, 238-249
FM- 108 O'Brien, D. A. and Herakovich, C. T.,
Finite Element Stress Analysis of Idealized Composite Damage Zones, NASA Rep CR-155923, Feb. 1978
FM- 109 Ouyang, C. and Lu, M.,
On A Micromechanical Model for Cracked Reinforced Composites, *Appl. Math. Mech.*, Vol. 2, No. 3, 1981, 297-304
FM- 110 Owen, M. J. and Bishio, P. T.,
Prediction of Static and Fatigue Damage and Crack Propagation in Composite Materials, Proc. AGARD Conf., Munich, Oct. 1974
FM- 111 Papaioannou, S. G.,
A Finite Element Method for Calculating Stress Intensity Factors and its Application to Composites, M.S. Thesis, Lehigh Univ., Mech. Engng., Bethlehem, 1971
FM- 112 Papaioannou, S. G. and Hilton, P. D.,
Finite Element Method for Calculating Stress Intensity Factors and Its Application to Composites, *Engng. Fract. Mech.*, Vol. 6, No. 4, 1974, 807-823
FM- 113 Parhizgar, S. and *et al.*,
Application of the Principles of Linear Fracture Mechanics to the Composite Materials, *Int. J. Fracture*, Vol. 20, No. 1, 1982, 3-15
FM- 114 Patil, S. S. and Rao, A. K.,
Special Finite Elements for Flaw and Discontinuity Stresses in Composites, 2nd Int. Conf. Str. Mech. Reactor Tech., Berlin, Sept. 1973
FM- 115 Ratwani, M. M.,
Characterization of Fatigue Crack Growth in Bonded Structures - Analysis of Cracked Bonded Structures, AFFDL-TR-77-31, Vol. II, 1977
FM- 116 Ratwani, M. M.,
Analysis of Cracked, Adhesively Bonded Laminated Structures, *AIAA J.*, Vol. 17, No. 9, 1979, 988-994
FM- 117 Ratwani, M. M. and Kan, H. P.,
Compression of Fatigue Analysis of Fiber Composites, *J. Aircraft*, Vol. 18, No. 6, 1980, 458-462
FM- 118 Repnau, T. and Adams, D. F.,
High-Performance Composite Materials for Vehicle Construction: A Finite Element Computer Program for the Elastoplastic Analysis of Crack Propagation, Rand Corp., R-1392-PR, Santa Monica, 1973
FM- 119 Rowlands, R. E. and Daniel, I. M.,
Stress and Failure Analysis of a Glass-Epoxy Composite Plate with a Circular Hole, *Exp. Mech.*, Vol. 13, No. 1, 1973, 31-37
FM- 120 Rybicki, E. F.,
An Energy Release Rate Approach for Stable Crack Growth in the Free-Edge Delamination Problem, *J. Comp. Mat.*, Vol. 11, 1977, 470-487
FM- 121 Rybicki, E. F. and Schmueser, D. W.,
Three-Dimensional Stress Analysis of a Laminated Plate Containing an Elliptical Cavity, Battelle Columbus Lab., AFML-TR-76-32, April 1976
FM- 122 Sih, G. C.,
Fracture Mechanics of Composite Materials, in Fract. Comp. Mater., Sijthoff & Nordh, 1979, 111-130
FM- 123 Smelser, R. E.,
Evaluation of Stress Intensity Factors for Bimaterial Bodies, *Int. J. Fract.*, Vol. 15, No. 2, 1979, 135-143
FM- 124 Staab, G. H.,
Finite Element Estimates of Singularity Powers for Cracks Terminating at Finite Width Interfaces, *Computers & Structures*, Vol. 18, No. 6, 1984, 1069-75

FM- 125 Staab, G. H. and Chang, T. C.,
A Finite Element Analysis of the Mixed Mode Bi-Material Fracture Mechanics
Problem, *Computers & Structures*, Vol. *18*, No. 5, 1984, 853-859
FM- 126 Steven, G. P.,
Finite Element Modelling of Fibre Pull-Out in Cracked Cement Composites, 4th
Int. Conf. Austral. FEM, Melbourne, Aug. 1982, 200-204
FM- 127 Sun, C. T. and Sierakowski, R. L.,
Fracture Characterization of Composites with Chopped Fiberglass Reinforcement,
SAMPE Quart., Vol. *11*, No. 4, 1980, 15-21
FM- 128 Tirosh, J.,
Effect of Plasticity and Crack Blunting on the Stress Distribution in
Orthotropic Composite Material, ASME 73-APMW-2, 1973
FM- 129 Ueng, C. E. S. and et al.,
Tensile Analysis of an Edge Notch in a Unidirectional Composite, *J. Comp. Mat.*,
Vol. *11*, 1977, 222-234
FM- 130 Walsh, P. F.,
Linear Fracture Mechanics in Orthotropic Materials, *Engng. Fract. Mech.*, Vol.
4, 1972, 533
FM- 131 Walsh, R. M.,
Strain Energy Release Rate Determination of Stress Intensity Factors by
Finite Element Methods, Center Compos. Mater., Univ. of Delaware, Rep. 82-06,
1982
FM- 132 Wang, A. S. D.,
Growth Mechanisms of Transverse Cracks and Ply Delamination in Composite
Laminates, Proc. ICCM-III, Paris, Vol. *I*, 1980, 170
FM- 133 Wang, A. S. D.,
Fracture Mechanics of Sub-Laminate Cracks in Composite Laminates, *Comp. Technol.
Review*, Vol. *6*, 1984, 45-62
FM- 134 Wang, A. S. D. and Crossman, F. W.,
Initiation and Growth of Transverse Cracks and Edge Delamination in Composite
Laminates. Part 1 An Energy Method, *J. Comp. Mat.*, Vol. *14*, 1980, 71-87
FM- 135 Wang, A. S. D. and Crossman, F. W.,
Fracture Mechanics of Sublaminate Cracks, Tech. Rep. AFOSR-TR-83-0594, 1983
FM- 136 Wang, A. S. D. and Crossman, F. W.,
Fracture Mechanics of Transverse Cracks and Edge Delamination in Graphite-
Epoxy Composite Laminates, Air Force Office Sci. Res., Final Tech. Rep.,
March 1982
FM- 137 Wang, A. S. D. and Law, G. E.,
An Energy Method for Multiple Transverse Cracks in Graphite-Epoxy Laminates,
Modern Dev. Comp. Mater. Struct., ASME, 1979, 17
FM- 138 Wang, A. S. D. and et al.,
Interlaminar Failure in Epoxy Based Composite Laminates, 29th Symp. Adv. Comp.
Design Appl., NBS, 1979, 255
FM- 139 Wang, A. S. D. and et al.,
On Mixed-Mode Fracture in Off-Axis Unidirectional Graphite Epoxy Composites,
Proc. ICCM-4, Tokyo, Japan, Vol. *1*, 1982, 599
FM- 140 Wang, S. S.,
Delamination Fracture from Surface Notch in 45Deg-/0Deg/90Deg Graphite/Epoxy
Composites, 2nd Int. Conf. Comp. Mat., TMS-AIME, 1978, 277-291
FM- 141 Wang, S. S.,
An Analysis of Tapered Double-Cantilever-Beam Fracture Toughness Test for
Adhesive Joints, in Fract. Mech., ASTM, STP 677, 1979, 651-667
FM- 142 Wang, S. S. and Yau, J. F.,
An Analysis of Cracks Emanating from a Circular Hole in Unidirectional Fibre
Reinforced Composites, *Engng. Fract. Mech.*, Vol. *13*, 1980, 57-67
FM- 143 Wang, S. S. and Yuan, F. G.,
A Hybrid Finite Element Approach to Composite Laminate Elasticity Problems
with Singularities, J. Appl. Mech., ASME, Vol. *50*, 1983, 835-844
FM- 144 Wang, S. S. and et al.,
Three-Dimensional Solution for a Through-Thickness Crack in a Cross-Plied
Laminate, Fract. Mech. Comp., ASTM STP 593, ASME, 1975, 36-60

FM- 145 Wang, S. S. and et al.,
A Multilayer Hybrid-Stress Finite Element Analysis of a Through-Thickness Edge Crack in a 45 Laminate, *Engng. Fract. Mech.*, Vol. *9*, 1977, 217-238

FM- 146 Wang, S. S. and et al.,
Fracture of Random Short-Fiber SMC Composite, *J. Comp. Mat.*, Vol. *17*, 1983, 299

FM- 147 Wang, S. S. and et al.,
Fracture of Adhesive Joints, Rep. R76-1, MIT, Dept. of Mater. Sci., Cambridge, 1976

FM- 148 Wang, S. S. and et al.,
A Mixed-Mode Crack Analysis of Rectilinear Anisotropic Solids Using Conservation Laws of Elasticity, *Int. J. Fract.*, Vol. *16*, No. 3, 1980, 247-259

FM- 149 Wen-Hwa, C. and Chei-Wei, W.,
On Elastodynamic Fracture Mechanics Analysis of Bi-Material Structures Using Finite Element Method, *Engng. Fract. Mech.*, Vol. *15*, 1-2, 1981, 155-168

FM- 150 Whitcomb, J. D.,
Experimental and Analytical Study of Fatigue Damage in Notched Graphite/Epoxy Laminates, NASA Tech. Mem. TM-80121, June 1979

FM- 151 Williams, R. S. and Reifsnider, K. L.,
Strain Energy Release Rate Method for Predicting Failure Modes in Composite Materials, Fracture Mech., ASTM STP 677, 1979, 629-650

FM- 152 Wolff, E. G. and et al.,
Opto-Acoustic Detection of Thermally Induced Microcracking in Al/CFRP Joints, *Composites*, Vol. *13*, No. 3, 1982, 323-328

FM- 153 Yamada, Y. and Okumura, H.,
Finite Element Analysis of Stress and Strain Singularity Eigenstate in Inhomogeneous Media or Composite Materials, in Hybrid Mixed FEM, S. N. Atluri, et al., Wiley, 1983, 325-343

FM- 154 Younan, M. Y. A. and et al.,
Crack Propagation Analysis for Orthotropic Non-Homogeneous Materials, *Engng. Fract. Mech.*, Vol. *16*, No. 2, 1982, 189-205

FM- 155 Zako, M. and Miyoshi, T.,
Fracture Simulation of Composite Material by Finite Element Method, 2nd Int. Conf. Comp. Mater., Toronto, Ontario, Apr. 1978, 622-634

FM- 156 Zako, M. and et al.,
Study on Fracture Toughness for Composite Materials, *Trans. Japan Soc. Mech. Eng.*, No. 395

FM- 157 Zhang, S. Y.,
A Study on Fracture Mechanism of Unidirectional Fibrous Composites, 4th Int. Conf. Comp. Mat., Tokyo, Oct. 1982, 617-624

FM- 158 Zhang, S. Y.,
Fracture Mechanics of Composites, *Progr. Mech.*, Vol. *10*, 2-3, 1980

Joints, Bonding (JB)

JB- 1 Aberson, J. A.,
Characterization of Crack Growth in Bonded Structure, 12th Ann. Meet. Soc. Engng. Sci., Univ. Texas, Austin, Oct. 1975

JB- 2 Adams, R. D. and et al.,
Prediction of Strength of Joints Between Composite Materials, Symp. Joint Fibre. Reinf. Plast., Imper. Coll., London, 1978

JB- 3 Agarwal, B. L.,
Static Strength Prediction of Bolted Joint in Composite Material, *AIAA J.*, Vol. *18*, No. 11, 1980, 1371-75

JB- 4 Allred, R. E. and Guess, T. R.,
Efficiency of Double-Lapped Composite Joints in Bending, Composites, Vol. *9*, No. 2, 1978, 112-118

JB- 5 Amijima, S. and et al.,
Stress Analysis and Strength of Adhesive Bonded Joints Under Bending Loads, 4th Int. Conf. Comp. Mat., Tokyo, 1982, 313-320

JB- 6 Amijima, S. and et al.,

Two Dimensional Stress Analysis of Adhesive Bonded Joints, 20th Japan Cong. Mater. Res., JSMS, 1977, 275-281
JB- 7 Amijima, S. and et al.,
Strength and Stress Analysis of Adhesive Bonded Joints of Isotropic and Anisotropic Materials, ICCM-II, 1978, 1185-99
JB- 8 Amijima, S. and et al.,
Stress Wave Propagation in Laminated Composites Having Visco-Elastic Adhesive Layers, ICCM-II, 1978, 1211-33
JB- 9 Amijima, S. and et al.,
A Micro-Computer Program for Stress Analysis of Adhesive Bonded Joints, J. Adhes. Soc. Japan, Vol. 19, No. 12, 1984, 561-459
JB- 10 Amijima, S. and et al.,
A Micro-Computer Program for Elasto-Plastic Stress Analysis of Adhesive Bonded Joints, The Sci. & Eng. Rev. Doshisha Univ., Vol. 24, No. 4, 1984
JB- 11 Anderson, G. P.,
Applied Adhesive Fracture Mechanics, Ph D Dissert., Univ. of Utah, Salt Lake City, Utah, 1973
JB- 12 Anderson, G. P.,
Analysis of Adhesive Bonds, Academic Press, New York, 1977
JB- 13 Anderson, G. P. and et al.,
Finite Element in Numerical Analysis of Adhesive Fracture, Int. J. Fract., Vol. 9, No. 3, 1973, 335-336
JB- 14 Anderson, G. P. and et al.,
Finite-Element in Adhesion Analyses, Int. J. Fract., Vol. 9, No. 4, 1973
JB- 15 Atluri, S. and Kathiresan, K.,
Analysis of Cracks in Adhesively Bonded Structures, ASME Winter Ann. Meet., Atlanta, Nov. 1977
JB- 16 Barker, R. M. and Hatt, F.,
Analysis of Bonded Joints in Vehicular Structures, AIAA J., Vol. 11, No. 12, 1973, 1650
JB- 17 Baumann, E.,
Finite Element Analysis of Advanced Composite Structures Containing Mechanically Fastened Joints, Nuclear Engng. Design, Vol. 70, No. 1, 1982, 67-83
JB- 18 Bennett, S. J. and et al.,
Adhesive Fracture Mechanics, Int. J. Fract., Vol. 10, No. 1, 1974, 33-43
JB- 19 Broutman, L. J. and Agarwal, B. D.,
Theoretical Study of the Effect of the Interface on Composite Toughness, 28th Annual Conf. Spi. Reinf. Plast. Div., Washington, D.C., 1973
JB- 20 Carpenter, W. C.,
Stress in Bonded Connections Using Finite Elements, Int. J. Num. Meth. Engng., Vol. 15, 1980, 1659-80
JB- 21 Carpenter, W. C.,
Finite Element Analysis of Bonded Connections, Int. J. Num. Meth. Engng., Vol. 6, 1973, 450-451
JB- 22 Chan, W. W. and Sun, C. R.,
Interfacial Stresses and Strength of Lap Joints, 21st Conf. Str., Str. Dyn. and Mater., Seattle, 1980
JB- 23 Chang, F. K.,
Bolted Joints in Laminated Composites, 25th Str., Str. Dyn. Mater. Conf., Palm Spring, May 1984, 252-256
JB- 24 Chang, F. K. and et al.,
Strength of Mechanically Fastened Composite Joints, J. Comp. Mat., Vol. 16, 1982, 470-494
JB- 25 Chang, F. K. and et al.,
Design of Bolted Composite Joints, 28th Nat. SAMPE Symp., April 1983, 164-175
JB- 26 Chang, F. K. and et al.,
Strength of Bolted Joints in Laminated Composites, AFWAL-TR-84-4029, Flight Dyn. Lab., July 1984
JB- 27 Chang, F. K. and et al.,
Strength of Mechanically Fastened Composite Joint, AFWAL-TR-82-9045, Flight Dyn. Lab., June 1982

JB- 28 Chen, W. H. and Wu, C. W.,
On the J-Integral for a Pressurized Crack in Bonded Materials, Int. J. *Fract.*,
Vol. *16*, No. 2, 1980, 47-51

JB- 29 Chen, W. H. and Wu, C. W.,
Finite Element Analysis of Bonded Materials With a Crack Perpendicular to the
Interface, 3rd Nat. Conf. Theor. Appl. Mech., ROC, 1979, 95-107

JB- 30 Chow, C. L. and *et al.*,
On the Determination and Application of COD to Epoxy-Bonded Aluminium Joints,
J. *Strain Anal.*, Vol. *14*, No. 2, 1979, 37-42

JB- 31 Cooper, P. A. and Sawyer, J. W.,
A Critical Examination of Stresses in an Elastic Single Lap Joint, NASA Rep.
TP-1507, 1979

JB- 32 Cope, R. D. and Pipes, R. B.,
Design of the Composite Spar-Wingskin Joint, *Composites*, Vol. *13*, No. 1, 1982

JB- 33 Coyle, E. J. and *et al.*,
Analysis of Cylindrical Joints of Composite Materials for Torsional Loadings,
J. *Reinf. Plast. Comp.*, Vol. *1*, No. 3, 1982, 195-205

JB- 34 Dattaguru, B. and *et al.*,
Geometrically Nonlinear Analysis of Adhesively Bonded Joints, NASA TM-84562,
Sept. 1982

JB- 35 Delale, F. and *et al.*,
Stresses in Adhesively Bonded Joints: A Closed-Form Solution, J. *Comp. Mat.*,
Vol. *15*, 1981, 249-271

JB- 36 Dickson, J. N. and *et al.*,
Development of an Understanding of the Fatigue Phenomena of Bonded and Bolted
Joints in Advanded Filamentary Composite Materials, WPAFB, Rep. AFFDL-TR-72-64,
Ohio, 1972

JB- 37 Everett, R. A.,
The Role of Peel Stresses in Cyclic Debonding, Adhesive Age, Vol. *26*, No. 5,
1983

JB- 38 Flaggs, D. L. and Crossman, F. W.,
Viscoelastic Response of a Bonded Joint Due to Transient Hygrothermal
Exposure, in Modern Dev. Comp. Mat. Str., ASME, New York, 1978, 299-314

JB- 39 Gali, S. and Ishai, O.,
The Time-Dependent Mechanical Behaviour of an Interlaminar Layer in a
Structural Bonded Model, J. Adhesion, Vol. *12*, No. 2, 1981, 113-125

JB- 40 Gillespie, J. W. and Pipes, R. B.,
Behaviour of Integral Composite Joints - Finite Element and Experimental
Evaluation, J. *Comp. Mat.*, Vol. *12*, 1978, 408-421

JB- 41 Gradin, P. A.,
A Fracture Criterion for Edge-Bonded Bimaterial Bodies, J. Comp. Mat., Vol. *16*,
1982, 448-456

JB- 42 Graves, S. R. and Adams, D. F.,
Analysis of a Bonded Joint in a Composite Tube Subjected to Torsion, J. Comp.
Mat., Vol. *15*, 1981, 211-224

JB- 43 Guess, T. R. and *et al.*,
Comparison of Lap Shear Test Specimens, J. Testing Evaluat., Vol. *5*, No. 3,
1977

JB- 44 Harrison, N. L. and Harrison, W. J.,
The Stresses in an Adhesive Layer, J. *Adhesion*, Vol. *3*, 1972, 195-212

JB- 45 Herakovich, C. T.,
On Thermal Edge Effects in Composite Laminates, Int. J. Mech. Sci., Vol. *18*,
1976, 129-134

JB- 46 Humphreys, E. A. and Herakovich, C. T.,
Nonlinear Analysis of Bonded Joints with Thermal Effects, NASA, CR-153263,
June 1977

JB- 47 Itani, R. Y,,
Elastic/Plastic Torsion of Non-Uniform and Non-Homogeneous Bars, Thesis,
Virginia Polyt. Inst. State Univ., Blacksburg, June 1975

JB- 48 Jones, R. and *et al.*,
Analysis of Bonded Repairs to Damaged Fibre Composite Structures, Engng. Fract.
Mech., Vol. *17*, No. 1, 1983, 37-46

JB- 49 Lakshminarayana, H. V.,
Finite Element Analysis of Laminated Composite Shell Junctions, *Computers & Structures*, Vol. *6*, 1976, 11-15
JB- 50 Lehman, G. M. and Hawley, A. V.,
Investigation of Joints in Advanced Fibrous Composites for Aircraft Structures, US Air Force, AFFDL-TR169-43, Vol. *1*, June 1979
JB- 51 Lene, F. and Leguillon, D.,
Effects of Slip Between Components in a Composite Material on its Effective Constitutive Coefficients, *J. de Mecanique*, Vol. *20*, No. 3, 1981, 509-536
JB- 52 Mall, S. and et al.,
Cyclic Debonding of Adhesively Bonded Composites, NASA TM-84577, Nov. 1982
JB- 53 Matthews, F. L. and et al.,
A Review of the Strength of Joints in Fibre-Reinforced Plastics, Part 2 Adhesively Bonded Joints, *Composites*, Vol. *13*, No. 1, 1982, 29-37
JB- 54 Matthews, F. L. and et al.,
Stress Distribution Around a Single Bolt in Fibre-Reinforced Plastic, *Composites*, Vol. *13*, No. 3, 1982, 316-322
JB- 55 Mitchell, R. A. and et al.,
Component Parts Assembly With Joints, Adhesive-Mechanical, US Air Force Rep. AFFDL-TR-169-43, Vol. *1*, June 1979
JB- 56 Mitchell, R. A. and et al.,
High Strength End Fitting for FRP Rod and Rope, J. Engng. Mech. Div., ASCE, Vol. *100*, 1974, 687-706
JB- 57 Nagaraja, Y. R. and Alwar, R. S.,
Nonlinear Stress Analysis of an Adhesive Tubular Lap Joint, J. Adhesion, Vol. *10*, No. 2, 1979, 97-106
JB- 58 Norwood, C. J. and Brown, K. C.,
Tubular Joint Design in Carbon Fibre-Reinforced Plastic, *J. Comp. Mat.*, Vol. *15*, 1981, 359-370
JB- 59 Norwood, C. J. and Brown, K. C.,
Finite Element Analysis of a Tubular Joint Design in Carbon Fibre Reinforced Plastic, 4th Int. Conf. Austral. FEM, Melbourne, Aug. 1982, 165-168
JB- 60 Pickett, A. and et al.,
Analysis of a Crimped and Bonded Joint for Load Bearing Skeletal Members, *Composites*, Vol. *13*, No. 3, 1982, 257-267
JB- 61 Raju, I. S. and et al.,
Two-Dimensional, Quasi Three-Dimensional and Three-Dimensional Analyses of Composite Joints, 5th ASCE-EMD Spec. Conf., Univ. Wyoming, Laramie, Aug. 1984
JB- 62 Ratwani, M. M.,
Characterization of Fatigue Crack Growth in Bonded Structures - Analysis of Cracked Bonded Structures, AFFDL-TR-77-31, Vol. *II*, 1977
JB- 63 Tirosh, J. and et al.,
The Role of Fibrous Reinforcements Well Bonded or Partially Debonded on the Transverse Strength of Composite Materials, *Engng. Fract. Mech.*, Vol. *12*, 1979, 267-277
JB- 64 Wang, S. S.,
An Analysis of Tapered Double-Cantilever-Beam Fracture Toughness Test for Adhesive Joints, in Fract. Mech., ASTM, STP 677, 1979, 651-667
JB- 65 Wang, S. S. and et al.,
Fracture of Adhesive Joints, Rep. R76-1, MIT, Dept. of Mater. Sci., Cambridge, 1976
JB- 66 Wang, S. S. and et al.,
Analysis of Lap Shear Adhesive Joints With and Without Short Edge Cracks, Rep. R76-2, MIT, Dept. of Mater. Sci., Cambridge, 1976
JB- 67 Wolff, E. G. and et al.,
Opto-Acoustic Detection of Thermally Induced Microcracking in Al/CFRP Joints, Composites, Vol. *13*, No. 3, 1982, 323-328
JB- 68 Wong, C. M. S. and Mathews, F. L.,
A Finite Element Analysis of Single and Two-Hole Bolted Joints in Fibre Reinforced Plastics, J. Comp. Mat., Vol. *15*, 1981, 481-491
JB- 69 Wooley, G. R. and Carver, D. R.,

Stress Concentration Factors for Bonded Lap Joints, *J. Aircraft*, Oct. 1971, 817
JB- 70 Wright, M. D.,
Stress Distribution in Carbon Fibre-Reinforced Plastic Joints, Composites, Vol. *11*, No. 1, 1980, 46-50
JB- 71 Wright, M. D.,
The Stress Analysis of a Butt Strap Joint in Carbon Fibre Reinforced Plastic, Composites, Vol. *9*, No. 4, 1978, 259-262
JB- 72 Yaniv, G. and Ishai, O.,
Residual Thermal Stresses in Bonded Metal and Fiber-Reinforced Plastic Systems, *Comp. Technol. Review*, Vol. *3*, No. 4, 1981, 131-137
JB- 73 York, J. L. and *et al.*,
Analysis of the Net Tension Failure Mode in Composite Bolted Joints, *J. Reinf. Plast. Comp.*, Vol. *1*, 1982, 141-152

CASE STUDY INDEX

The following case study index only includes the industrial examples described by the authors in their papers. Most of the programs have been used worldwide to solve a broader range of industrial problems. However, the results of such investigations are not always readily available for publication as they remain the property of the users. The present non-exhaustive study index should therefore not be considered by the reader as an indication of a program's capability, but only as a subject index.

Actuator arm, FIESTA
Aculator carriage, ANSYS
Air systems, CASTOR
Aircraft wheel leg, SAMKE
Arch, ELASTODYNAMICS (2D)
   bridge arch, AFAG, RECAFAG
   concrete, ASE
   dam, FIESTA, MODULEF, ZERO-4
Automobile body structure, ALSA

Bar, SURFOPT
Beam, S AND CM PACKAGE
Birfurcated duct, FIESTA
Bolt, ANSYS, SURFOPT, TITUS
Bracket, FIESTA
Branched structure, BOSOR 4
Bridge arch, AFAG, RCAFAG
Building, FLASH

Casks, FEMFAM
Chip/chip carrier, FEMPAC
Chopper, FEMFAM
Church, FIESTA
Coke oven, INFESA
Compressor casing, CASTOR
Concrete
   arch dam, ASE
   plate, KYOKAI
   slab, ADINA, ASE
   wall, INFESA

Connecting
   flange, CASTOR
   joint, MEF/MOSAIC
   rod, BEASY
Containment vessel, PANDA
Continuous beam, S AND CM
Cooling
   hole, FIESTA
   tower, LASSAQ
   water, AXISYMMETRIC PACKAGE
Crane, FLASH
Crank arm, CASTOR
Crankshaft, BEASY
Cryogenic cooler, BOSOR 4
Cylinders, CASTEM
   ring-stiffened, BOSOR 5
Cylindrical
   panel, PANDA
   shell, BOSOR 5, LASSAQ

Dam, FIESTA
   arch, FIESTA, MODULEF, ZERO-4
   concrete, ASE
   foundation, FIESTA
Domes, AXISYMMETRIC PACKAGE, STDYNL, THERMAL PACKAGE

Earthquake, PAID, TITUS
Electric engine, MODULEF
Ellipsoidal tank, BOSOR 4
Excavator, FEMPAC

Case Study Index

Fibre reinforced plastics, LASSAQ
Floating frame, FENRIS
Floor panel, ALSA
Fluid structure, ADINA,
  AXISYMMETRIC PACKAGE, MODULEF,
  TITUS
Flywheel, ROBOT
Food processor, PDA/PATRAN
Foundation, OSTIN
Fracture mechanics, CASTEM, TITUS
Frame, AFAG, DEFOR, FENRIS, RCAFAG,
  S AND CM, STDYNL, THERMAL PACKAGE

Gear, CASTOR, UNIC GEAR
  case, FIESTA

Heat
  exchanger, ROBOT
  generator, CASTEM
Hexagonal bundle,, KYOKAI
Housing, AIT, FEMFAM
Human femur bone, FIESTA, MODULEF
Hydraulic engine, STRUGEN

Imperfect cylinder, CASTEM

Jacket, FENRIS

Landing gear, FIESTA

Mast antenna, REST
Mining excavator, NE-XX
Missile impact, CASTEM
Mixing drum, HYBRID
Motorway bridge, MICRO-STRESS, NE-XX

Notched parts, CASTOR
Nozzle, BEWAVE, CASTEM, FIESTA,
  PDA/PATRAN, TITUS
Nuclear reactor, PANDA
  housing, ZERO-4

Offshore, AQUADYN, STDYNL
  flexible arch, FLEXAN
Outlet nozzle, CASTEM

Parabolic dome, STDYNL
Pipe, ADINA, PAID
  impact, CASTEM
Pipework system, THERMAL PACKAGE
Piston, FIESTA, PDA/PATRAN
Plane frame, DEFOR, S AND CM
Plate, BEWAVE, FEMFAM, HYBRID, SAMKE
  with circular hole, NE-XX
  with variable section, ESA
Portal frame, AFAG, RCAFAG
Pressure vessel, CASTOR, MEF/MOSAIC
  head, BOSOR 5
Pylon, MICRO-STRESS

Railway wagon, FEMFAM
Reactor vessel, RAPS
  shroud, BOSOR 4

Rigid-jointed frame, S AND CM
Ring stiffener, AXISYMMETRIC PACKAGE
Ring-stiffened cylinder, BOSOR 5
Rockets, BOSOR 4, BOSOR 5, PDA/PATRAN
Rotor, MODULEF
  disk, MEF/MOSAIC

Satellite structure, SIMP
Shearer arm, PAFEC
Shell structure, CASTEM
Shock absorber, MEF/MOSAIC
Silo construction, FEMPAC
Skew grid, S AND CM
Soil structure, OSTIN
Solar arrays, SIMP
  cell, SIMP
Space frame, DEFOR, THERMAL PACKAGE
Spherical dome, STDYNL
Spray nozzle, TITUS
Stator, MODULEF
Statue of Liberty, CASTOR
Steam generator, BEWAVE, TITUS
Steel structure, ESA
Stiffened torus, CASTEM
Structural steel work, STAR 2
Submarine finder, CASTEM
Support rail, ALSA
Suspended bridge, TITUS
Syphon tank, ROBOT

Tank, ESA
Tapered disc, AXISYMMETRIC PACKAGE
Tennis racket, DAPST
Tension tower, DEFOR
Thin-walled cylinder, AXISYMMETRIC
  PACKAGE
Three dimensional frame, IBA
Torus, CASTEM
Trolley reinforcement, SIMP
Truss, S AND CM
Tube, MSRC-RB
Tubular joint, SESAM-80
Tunnel, ADINA,
  ELASTODYNAMICS (2D), OSTIN
Turbines, ANSYS, CASTOR, FLASH,
  PDA/PATRAN
Turbo alternator, FIESTA
Turbo-jet, TITUS

Valves, CASTOR, FEMFAM, FIESTA,
  PDA/PATRAN

Wall, ESA
Water
  injection platform, SESAM-80
  pipe, ADINA
Watertank, BOSOR 5, CASTOR

X-braced frame, FENRIS
X-ray tube, ANSYS

RAYMOND H. FOGLER LIBRARY

DATE DUE

BOOKS ARE SUBJECT TO
RECALL AFTER TWO WEEKS.